WHAT COLOUR IS THE SUN?

Also by Brian Clegg
Dice World
How Many Moons Does the Earth Have?
Inflight Science
Introducing Infinity: A Graphic Guide
Light Years
Science for Life
The Quantum Age
The Universe Inside You

WHAT COLOUR IS THE SUN?

MIND-BENDING SCIENCE FACTS IN THE SOLAR SYSTEM'S **BRIGHTEST** QUIZ

BRIAN CLEGG

ICON

Published in the UK in 2016 by
Icon Books Ltd, Omnibus Business Centre,
39–41 North Road, London N7 9DP
email: info@iconbooks.com
www.iconbooks.com

Sold in the UK, Europe and Asia
by Faber & Faber Ltd, Bloomsbury House,
74–77 Great Russell Street,
London WC1B 3DA or their agents

Distributed in the UK, Europe and Asia
by Grantham Book Services,
Trent Road, Grantham NG31 7XQ

Distributed in the USA
by Publishers Group West,
1700 Fourth Street, Berkeley, CA 94710

Distributed in Australia and New Zealand
by Allen & Unwin Pty Ltd,
PO Box 8500, 83 Alexander Street,
Crows Nest, NSW 2065

Distributed in South Africa by
Jonathan Ball, Office B4, The District,
41 Sir Lowry Road, Woodstock 7925

Distributed in India by Penguin Books India,
7th Floor, Infinity Tower – C, DLF Cyber City,
Gurgaon 122002, Haryana

Distributed in Canada by Publishers Group Canada,
76 Stafford Street, Unit 300
Toronto, Ontario M6J 2S1

ISBN: 978-178578-149-0 (US edition)
ISBN: 978-178578-122-3 (International edition)

Text copyright © 2016 Brian Clegg

Typeset in PMN Caecilia by Marie Doherty

Printed and bound in the UK
by Clays Ltd, St Ives plc

For Gillian, Rebecca and Chelsea

ABOUT THE AUTHOR

Science writer Brian Clegg studied physics at Cambridge University and specialises in making the strangest aspects of the universe – from infinity to time travel and quantum theory – accessible to the general reader. He is editor of www.popularscience.co.uk and a Fellow of the Royal Society of Arts. His previous books include *How Many Moons Does the Earth Have?*, *Ten Billion Tomorrows*, *Science for Life*, *Light Years*, *Inflight Science*, *Build Your Own Time Machine*, *The Universe inside You*, *Dice World*, *The Quantum Age* and *Introducing Infinity: A Graphic Guide*.

www.brianclegg.net

ACKNOWLEDGEMENTS

With many thanks to my editor Duncan Heath, and everyone at the excellent Icon Books for their support. And with particular thanks to everyone who has said nice things about this book's predecessor, *How Many Moons Does the Earth Have?*

CONTENTS

INTRODUCTION

What Colour Is the Sun? has a traditional quiz format. The book contains two quizzes, each with six rounds of eight questions, plus two themed 'special rounds' each offering up to ten points.

The chances are, though, that you will just enjoy the quiz by testing yourself, so the book is designed to be read through solo. Each answer is accompanied by illuminating information, so there is more to it than just getting the answer right. Of course, if you're using the book as a pub quiz, you don't need to include these parts.

If you are going to use the book in a real quiz, just copy the questions from the two special rounds and print out enough so that each team can have their own question sheet. You might like to use one of these as a 'table' round, which is left on the teams' tables to answer between the other rounds.

A popular addition in quiz play is to allow each team to have a joker to use on a round of their choice (before they see the questions), which doubles their points in that round.

The little 'while you're thinking' factoids after each question are primarily for your enjoyment, but depending on your audience, it might add to the fun to read them out when running a quiz.

Whichever way you use the book – enjoy it!

QUIZ 1
ROUND 1: BIOLOGY

Wrinkly extremities

Why do hands and feet go wrinkly in the bath?

Answer overleaf ➔

While you're thinking ...

The ideal bathwater temperature is generally agreed to be
around 38°C (100°F) – just over body temperature.

The skin on hands and feet reacts quite differently in
a bath than does skin elsewhere on the body.

Skin is waterproof, yet surprisingly it wasn't until 2012 that it
was discovered how the waterproofing worked. It's down to
a layer of fat molecules called lipids with two water-repelling
tails. Usually these molecules have the tails pointing in the same
direction, but in the skin they point in opposite directions and
are stacked together in a way that maximises waterproofing.

To give a better grip in the wet

We've all seen how the skin on our hands and feet goes wrinkly in the bath, and it's common to assume that this is because the skin takes in water – but the hands and feet are just as waterproof as the rest of the body. Instead this is a nervous system reflex that scientists speculate is to give a better grip in wet conditions.

It has been known for 80 years that the wrinkling is not about taking on water. One key indicator is that it doesn't take place if there is nerve damage in the locations where the skin wrinkling takes place, indicating that it is an action of the central nervous system. But it was only in 2011 that it was suggested that this reflexive action has a use – that it may have evolved because it proved valuable.

This was because wrinkly fingers and toes do have a practical benefit. They act like the indentations in a car tyre. Tyres have the best grip in the dry if they are slick, without any tread. But in the wet, the contoured surface of a road tyre provides channels for water to be taken away from the interface between the tyre and the tarmac, improving grip. Similarly, the wrinkles in our hands and feet work effectively to carry away water that would otherwise reduce our grip in the wet.

This hypothesis was tested by getting participants to pick up objects like marbles in both dry and wet conditions. When the hands were wet and had gained their wrinkling, they were better at picking up wet marbles than dry, wrinkle-free hands, but there was no difference when picking up dry marbles. The wrinkles, it seems, are human tyre treads for hand grip and to make it less likely we will fall over on wet surfaces.

Further reading: **Nature Wrinkly**

Asparagus pee

Some people don't find their urine smelly after eating asparagus. Is this because they don't produce the relevant chemical, or because they can't smell it?

Answer overleaf ➔

While you're thinking ...

We smell something when chemical molecules lock into olfactory receptors – cells at the back of the nose that have special proteins on their tips to accommodate specific molecular forms.

A human has around 450 olfactory receptors, about half the number that a dog has in its nose.

Women are generally better in smell-based tests than men. It has been suggested that this might be due to their having more cells (typically over 40 per cent more) in a region of the brain called the olfactory bulb, which is involved in processing the sense of smell.

It seems to be both (probably)

It has long been known that eating asparagus makes wee smell strange – but not why some people can't detect the odour. There is still dispute over the detail, with some scientists claiming that everyone produces the distinctively scented urine after eating and that the variation comes entirely from whether or not our noses can detect it. But it seems likely that some of us don't produce the chemical cocktail responsible for the smell and that others can't detect it, even when it is definitely present. It doesn't help that we aren't totally sure which chemicals are behind the odour, though a compound called methanethiol is a prime suspect.

One of the reasons there is a dispute is that there have been significant variations in the results studies have produced, with the percentage who *produce* the strange-smelling urine found to be around 50 per cent in some studies and as much as 90 per cent in others. (It has been suggested that the variation may reflect a regional genetic variation.) Similarly, tests looking for the ability to detect the distinctive smell have produced results varying from 10 per cent to practically the entire population.

It's not unusual for different studies to produce different results. Whenever a scientific test depends on humans describing a response, there is inevitably inaccuracy – and there may be genetic variations in populations around the world. But it is unusual to find such a distinctive split on whether or not something is happening at all. Asparagus is yet to fully give up its secrets.

Further reading: **BBC Future – Asparagus**

Does this bug you?

Is a bug a beetle?

Answer overleaf ➔

While you're thinking ...

New beetles are being found all the time. In 2016, 24 new species
of weevil were discovered in a single Australian collection alone.

The original Volkswagen Beetle car (properly the 'Volkswagen
Typ 1') was often known as the Volkswagen Bug in the USA.

The ladybird, known as a ladybug in the USA, is a beetle.

No, a bug is not a beetle

Although the term 'bug' is used very loosely for practically any creepy-crawly (even as a term for a bacterial or viral infection), it actually has a very specific meaning. Both bugs and beetles are insects, but bugs are of the order Hemiptera and beetles are Coleoptera.

Practically speaking, there are two big differences between the two – in the mouth and the wings. A bug has a beak-like sucker of a mouth, ideal to suck the juices out of a plant – aphids, for instance, are bugs – or to suck dry a prey insect (or, in something like a bed bug, which is a true bug, to extract the blood from a human). By contrast, a beetle has a more conventional 'chewing' mouth for dealing with solid foods.

As for the wings, if bugs have wings (some don't), they are fairly conventional membranes. But one of a beetle's two pairs of wings has been adapted by evolution to become a protective outer casing, so that when the insect isn't flying, the delicate proper wings are covered by the pair of 'elytra', as the casing wings are known.

There are far more types of beetle than of bug, with fewer than 100,000 types of bug identified to date, compared with well over 400,000 types of beetle – around a quarter of all known animals. It has been speculated the total number could be in the tens of millions. The biologist J.B.S. Haldane wrote 'The Creator would appear as endowed with a passion for stars, on the one hand, and for beetles on the other'. (Haldane is often quoted as saying the Creator had 'an inordinate passion for beetles', but this seems to be apocryphal.)

*Further reading: **The Collins Field Guide to Insects***

Twenty-one limb salute

What kind of animal (undamaged and intact) can have 21 limbs?

Answer overleaf ➜

While you're thinking ...

The formal definition of a limb is a part of the body separate from the head or trunk (or various other body parts depending on the kind of animal), so includes, for instance, arms, legs, wings and tentacles.

Centipedes can have around 30 legs (or many more), so could have 21 limbs by losing a few legs – but always have an even number when intact.

The majority of higher animals have an even number of legs – for example mammals have four, insects six and arachnids eight.

A starfish

Although a typical starfish has five limbs (usually called 'arms' because in locomotion they pull themselves along, rather than walk), the Crown of Thorns starfish (*Acanthaster planci*) can have as many 21 arms. Starfish are unusual in commonly having an odd number of limbs, which they can regrow if lost. Confusingly, while the starfish's limbs are called arms, the multiple projections from the underside of the arms are called tube feet.

Some argue that a kangaroo has five limbs, as the tail is used in locomotion, but this stretches a point – it is still a very different kind of limb from arms or legs.

The Crown of Thorns isn't the champion arm grower of the starfish world – the Antarctic species *Labidiaster annulatus*, for example, can have as many as 40–50 limbs, described as rays, which look like very slim tentacles. Like a number of simpler animals, the starfish has radial symmetry – its body is symmetrical on rotation – rather than the bilateral symmetry (mirror symmetry of two sides), typical in more sophisticated animals from cockroaches to humans.

One of the starfish's more impressive (if arguably some-what disgusting) capabilities is to be able to digest food outside its body. It can extrude a section of its stomach, for instance inserting it between the shells of a bivalve mollusc. The stomach starts digesting the food externally, and then is withdrawn with the remainder of the food into the starfish's body when the prey is broken down enough to be ingested.

Further reading: **Invertebrates**

Live and let live

Is a virus alive?

Answer overleaf ➜

While you're thinking ...

School biology generally defines life as requiring
seven processes: movement, nutrition, respiration,
excretion, reproduction, sensing and growth.

Many viruses are among the smallest things that could be
considered an organism, just 18–300 nanometres across.

Not all viruses attack humans, or even complex
life – many prey on bacteria.

Yes and no: a virus is and isn't alive

Award yourself a point for 'yes' and one for 'no', or two points for 'yes and no', 'we don't know' or 'it depends how you define "alive"'. There is no true agreement on whether or not a virus is alive. If you stick to the seven processes listed on the previous page, a virus isn't alive because, for example, it doesn't have a traditional metabolism. Neither does it have a built-in mechanism for reproduction; rather it hijacks the reproductive mechanism of another cell. In effect, a virus is a delivery mechanism for deoxyribonucleic acid (DNA) and ribonucleic acid (RNA) in order to reproduce and thrive.

When viruses were first discovered, they were thought to be simply smaller equivalents of bacteria, but once their reproduction method was identified, it became traditional to consider a virus to be not alive. However, a number of modern biologists question this argument, as the 'process of life' approach has severe limitations when working at the level of cells – for example, living human cells do not qualify as being alive if you apply the seven processes test to them.

Another argument against the dismissal of viruses from life is the discovery of extremely large mimiviruses which have more of the internal mechanisms that we would expect in a living organism than does a typical virus. And as recently as 2015 it was discovered by investigating genetic history that viruses share a range of properties with living cells and probably originated as more traditional cellular organisms which then shed complexity as an evolutionary benefit, rather than having always been so simple.

Overall, the jury is still out, but the evidence has swung somewhat towards viruses being alive.

*Further reading: **Reality's Frame***

Beginning with a flash in the pan

What is panspermia?

Answer overleaf ➜

While you're thinking ...

Pan was the name of an Ancient Greek god, responsible for the mountains, wild areas, shepherds and their flocks, and fertility.

English words beginning with 'pan' are borrowing from the Greek word πᾶν, the neuter form of the word for 'all'. It was widely used as a combined form in Greek words, such as panagia (all-holy) and panselene (of the full moon).

The 'spermia' part of the word is derived via Latin from the Ancient Greek verb 'to sow'.

The theory that all life on Earth was seeded from space

Panspermia was originally defined as the theory that life derives from tiny germs, which supposedly spread through the air and started to grow when they met a suitable location. By the start of the 20th century, however, it had come to mean the idea that life did not originate on Earth, but arrived, probably on meteors and comets, from somewhere else in the universe. (The idea itself can be traced back to the Ancient Greeks, but the term was applied to it only in the last century.)

It's fair to say that this has never been a mainstream theory, but it has had some notable supporters including Francis Crick, Fred Hoyle and Chandra Wickramasinghe. Hoyle and Wickramasinghe pointed out that a considerable amount of dust in space is organic, that it is possible for organic matter to exist as spores in space and to survive landing on Earth, and that life seems to have emerged around the time of the Late Heavy Bombardment when the Earth had an unusually high rate of asteroid impacts.

One of the attractions of panspermia is that it fits well with the difficulty of creating life and the way that all life appears to have a common ancestor – Hoyle and Wickramasinghe also suggested that panspermia could be responsible for the introduction of new diseases. However, most biologists are doubtful, arguing that panspermia adds an unnecessary complication and that in dealing with the difficulty of creating life, it merely pushes the boundary back to a different point in time and space.

*Further reading: **Reality's Frame***

What was I doing?

How long can a goldfish remember things?

Answer overleaf ➜

While you're thinking ...

A goldfish joke: 'Just because I have a [insert figure of your choice]-second memory, they don't think I'll mind eating the same fish flakes over and over ... Oh, brilliant! Fish flakes!'

The goldfish is a member of the carp family.

The basic carp species that the goldfish was bred from is silver/grey, but can be orange or red-coloured due to a mutation.

Goldfish remember things for several months at least

There is a common myth that goldfish have only a three-second memory span, hence the joke on the previous page. It's not clear where this myth came from, but some have suggested that it originated with an advertisement (rarely a good source for science). Remarkably, the myth has become so strong that it has appeared in some scientific papers on the behaviour of fish (admittedly usually to contradict it).

There are clear practical examples, familiar to any goldfish keeper, that demonstrate the fishes' memory working for longer than a few seconds. Goldfish, for example, learn to associate the rattling sound of a container of fish food with eating and will congregate, ready to be fed. There is also good evidence that goldfish have a degree of face recognition ability, and will often get more excited when their owner comes near (on the assumption that they will be fed) than a stranger.

The US Discovery Channel TV show *Myth Busters* took on the 'three second' myth in 2003 (broadcast in 2004) and demonstrated that a goldfish could recognise colour patterns and remember the route through a maze a month after learning it. In 2009, a Plymouth University team showed that goldfish could remember feeding prompts for at least three months.

Every few years, newspapers feature someone disproving the myth as if it had not been done before, suggesting, perhaps, that journalists have shorter memories than normal human beings.

Further reading: **Learning in Fishes**

Zircons are a scientist's best friend

Forget diamonds: why do biologists love zircons when it comes to establishing the age of life on Earth?

Answer overleaf ➜

While you're thinking ...

Zircon is a mineral that is primarily made up of zirconium silicate.

Although zircons are crystals, they tend to be small (mostly the size of sand grains) and are typically yellow, though they can be red, green, blue or colourless. Large colourless crystals are sometimes used in jewellery as a substitute for diamonds.

Although zircons are sometimes used in place of diamonds, they are a different compound to cubic zirconia – another, more popular faux diamond – which is a crystalline form of zirconium dioxide.

Zircons lock in small amounts of carbon, which can be used to infer the existence of early life

Zircons tend to hold impurities, including carbon, locked in when the crystals were formed, which can give an indication of the conditions on Earth at that time. The material is dated using uranium decay dating and results have suggested that some examples date from the early years after the formation of the Earth.

The presence of carbon is not, in itself, a definitive indicator of life; the carbon had to be here on Earth for the life to start from in the first place. But the interesting thing is the proportion of the different carbon isotopes that were locked away. Isotopes are variants of an atom with alternative numbers of neutrons in the nucleus. The most common version is carbon-12, with six protons and six neutrons. There is also the radioactive carbon-14 used in radiocarbon dating, with eight neutrons – but that technique is only useful for a few thousand years.

However, carbon comes in a third isotope, carbon-13, which has seven neutrons and is pretty much stable. The mechanisms that living organisms use to build up their cells have a preference for the smaller carbon-12 atoms, and so an accumulation of carbon from a living creature has slightly more carbon-12 than would be found in carbon that is simply part of a mineral deposit.

Zircons carrying the tell-tale extra carbon-12 have been found that are around 4 billion years old, making it seem likely that life had already started when the Earth was just 400–500 million years old. Which is why biologists love a zircon.

*Further reading: **Reality's Frame***

QUIZ 1
ROUND 2:
HISTORY OF SCIENCE

Walking the Planck

What subject did physicist Max Planck's physics professor advise him to study at university?

Answer overleaf ➜

While you're thinking ...

Max Planck was born in Kiel, Germany, in 1858.

The professor in question was Philipp von Jolly,
based at the University of Munich.

Max Planck would go on to take the essential theoretical step that
precipitated the transformation of physics by quantum theory.

Music

When the young Max Planck was exploring his options for university, he could, in principle, have chosen physics or music. While he had been both interested in physics and good at the subject, along with the underlying mathematics, he was also an excellent musician, playing the piano to concert standard. Planck genuinely could have opted for a career in music, and the physics professor, Philipp von Jolly, tried to persuade the young student that there really was not a lot of point in going with physics.

According to von Jolly, physics was pretty well finished as a discipline. He pointed out that apart from a few small gaps, almost everything that needed to be discovered already had been. It seemed that von Jolly was suggesting that there was no opportunity for Planck to cover himself in glory by discovering something new and exciting. However, von Jolly reckoned without Planck's modesty. The student said that he was not particularly interested in discovering new things, simply wishing to understand the existing physical theories.

We don't know how Planck would have turned out as a musician, but in physics he ended up outshining von Jolly and becoming such a name in German science that when it was considered politic to rename the society responsible for German science research from the Kaiser Wilhelm Society, it became the Max Planck Society, with the associated Institutes, now numbering over 80, known as the Max Planck Institutes.

It is ironic that, while von Jolly indeed did not add much that was new to the constituents of physics, Planck would help found quantum theory, one of two new areas of physics (the other being relativity) that blew apart von Jolly's comfortable picture of a near-complete knowledge of the subject.

*Further reading: **The Quantum Age***

What's in a name?

Where did the German natural philosopher Regiomontanus come from? (The clue's in the name.)

Answer overleaf ➜

While you're thinking ...

Some of the names by which the early scientists and philosophers are known are really Latinised versions of their name, or even of nicknames.

Some scientific nicknames referred to personal qualities, or to where the individual originated.

An example of a straightforward Latinised name would be the man who devised the biological naming system still used, Linnaeus, whose actual name was Carl von Linné, while the mathematician Tartaglia was actually called Niccolò Fontana, with Tartaglia being a nickname based on his stammering.

Regiomontanus came from Königsberg

It helps to know a little German to see how Johannes Müller could end up better known as Regiomontanus. The 15th-century mathematician and astronomer was born near Königsberg in what is now Bavaria. The name of the German town simply means King's Mountain – if there were an English equivalent, it would be something like Mountroy. And a Latinised version of King's Mountain would be Regiomontanus.

The tendency to give these Latin epithet names had some apparent similarities to the medieval tendency to use scholastic accolades, although the reason for doing so was rather different. Scholastic accolades were honorary titles, applied to theologians and other academics to imply that they were the outstanding scholars of their day. So, for instance, Thomas Aquinas was Doctor Angelicus and Roger Bacon was Doctor Mirabilis.

As almost all academic writing was in Latin all the way through to the 1600s – Isaac Newton, for instance, wrote his key work, *The Principia*, in Latin, though Galileo had already bucked the trend a few decades earlier by writing his books in Italian – it was common to use Latinised versions of names, which were usually simply achieved by sticking a Latin ending on to the surname, but occasionally, as with Regiomontanus, involved a more convoluted Latin equivalent to a nickname or birthplace.

Further reading: **The Quantum Age**

No half measures

What is the unit of measure the cubit based on?

Answer overleaf →

While you're thinking ...

Units of measurement are often based on a natural quantity or size.

The kilogram was originally defined as the weight of a cubic decimetre of distilled water at the temperature of melting ice, though for practical purposes, chunks of a platinum-iridium alloy of this mass were used as reference kilograms.

Perhaps the most obvious natural unit of measurement is the foot, though few individuals have a foot that is as long as a modern foot length, which is roughly the equivalent of a UK shoe size 13 (US size 14, EU size 47).

The distance from the elbow to the tip of the middle finger

The cubit is a common mid-range measure from ancient times. By using a part of the body as the standard, the cubit was easy to measure out without the need for any equipment. There are other body-based measures in the traditional system, though all were thrown away with the move to metric. So, for example, although the English word 'inch' comes from the Latin for one-twelfth, in many languages it corresponds to the word for thumb, as it was originally approximately the width of an adult male's thumb.

The cubit was renamed the ell in medieval times and was widely used in Europe as a measurement of cloth, for obvious reasons, as the arm could be used as an easy measure.

The trouble with all such person-based measures is that people come in a range of shapes and sizes. This makes the measures fine for approximate concepts, such as spacing out crops, but not when selling a product or doing mathematics, where a standard tends to be introduced.

Even when there was a standard, though – the ell, for example, was measured in England using something known as the ell-wand – that standard tended to vary from place to place, leading to potential confusion. So, for example, Archimedes, in his book *The Sand Reckoner*, makes an estimate of the size of the universe measured in stades – the distance around a Greek running track. But there was no fixed length for this: stades differed from city to city; so we aren't sure exactly what distance he had in mind.

*Further reading: **Measure for Measure***

Adding with letters

In Greek maths, the numbers were represented by letters ($\alpha = 1$, $\beta = 2$, $\gamma = 3$ ...) and fractions were represented by dashed letters ($\gamma' = \frac{1}{3}$, $\delta' = \frac{1}{4}$ ...). So what was β'?

Answer overleaf ➔

While you're thinking ...

To confuse things more, the Ancient Greeks used a couple of obsolete letters – such as the digamma, which had been dropped from the alphabet – to represent numbers.

The biggest number in the basic Greek system was a myriad (10,000), though a myriad myriads (100 million) was sometimes used.

The letter-based approach wasn't the only number system used by the Greeks, who started with more of a tally-based system. Letters are often still used in Greek for ordinal numbers – the numbers based on order, such as the third item in a list.

Two-thirds

Bizarrely, while almost any number with a dash was 1/that number, β′ was ⅔. The Ancient Greeks used a special character for ½, which looked like a zig-zag lightning bolt: ⚡.

Even more so than the more familiar Roman numbering system, the Greek system did not make arithmetic easy. It was bad enough adding two numbers when each number was represented by a different letter, but was even more complicated with fractions. Unable to apply the kind of top and bottom logic we use in handling fractions, the Greeks had to resort to published tables that contained the results of the tedious construction of different calculations.

It also didn't help that the Greeks weren't really thinking of fractions in the same way that we do. All their 'fractions' had 1 on the top. To get what we would call ⅗, for instance, they would have to use three lots of ⅕. This reflects that a quantity like γ′ was not seen as one-third, but as the third part. That sounds like simply a distinction of wording, but it's not. Rather than thinking of a whole number divided by three (hence a fraction), the Greeks were thinking of a whole number, three of which would make a bigger whole. It was a very visual approach, like thinking of three small stones making up a bigger stone. The mathematics works, but there are no fractional stones – they are all whole stones – and the Greek 'fractions' were conceptually whole numbers.

*Further reading: **A Brief History of Infinity***

Isaac's tomes

On which subject did Newton own the most books?

Answer overleaf ➔

While you're thinking ...

Isaac Newton had a very large library for the
time with around 2,100 titles.

After remaining pretty much in one piece through to the 20th century,
much of Newton's library was auctioned off in the 1920s.

To put the size of Newton's library into context, his Cambridge
college at the time, Trinity, which had one of the better
libraries, owned between 3,000 and 4,000 books.

Theology

You might expect that physics would be Isaac Newton's number one reading subject, but with 109 titles it only came in fifth place, after theology (477), fiction (207), alchemy (138) and mathematics (126), though admittedly the dividing line between physics and maths was not the same then as it is now – Newton's own masterpiece of physics, *The Principia*, was classified in his library under mathematics.

The large numbers of alchemy and theology books should be no great surprise, as Newton probably spent more time on these subjects than he did on science. Although not a conventional Anglican, he had strong Christian beliefs and spent a lot of time on theology and on detailed analysis of the Bible (working out, for instance, when he expected Christ's second coming to occur). Similarly, alchemy, which at the time had not entirely separated from chemistry, was a field in which he probably experimented more than he did in physics, though this activity was kept quiet, as alchemy was considered a little too close to magic to be respectable. Alchemy was divided into two, with some, like the early chemist Robert Boyle, concentrating primarily on speculative alchemy, which was closer to chemistry in studying how elements combined. But Newton seems to have been more interested in operative alchemy, which included the attempt to transmute base metals into gold.

Asked to think of categories included in Newton's library, it is easy to overlook those fiction titles because Newton is always portrayed as intensely serious and single-minded, but clearly he enjoyed fiction. Other subjects covered included economics and medals, reflecting his time at the Royal Mint.

*Further reading: **The Library of Isaac Newton***

QUESTION 6
Greek gyrations

The Ancient Greeks had a word for people who attempted a geometric feat we now know to be impossible: τετραγονιδζειν (tetragonidzein). What were those people trying to do?

Answer overleaf ➔

While you're thinking ...

Tetragonidzein were attempting to use geometry for a particular purpose. Geometry means 'measuring the Earth' in the original Greek.

Although the most famous geometric proof is called Pythagoras' theorem, it was practical in use long before Pythagoras lived and there is no evidence that he developed the formal proof.

If you were taught traditional geometry at school, you would have ended a proof with Q.E.D., standing for *quad erat demonstrandum*, roughly 'what was shown'. The original Greek version was *OEΔ*, which stands for a Greek phrase that was approximately 'the thing that was to be proved'.

Tetragonidzein were attempting to square the circle

Greek geometry existed in a strange detached world that had only a loose connection to reality. It was strongly influenced by ideas like that of Plato that there was an absolute archetype for an item like a triangle, and that what we saw in the world were just flawed versions, rather like shadows of an outside world cast on to the wall of a cave. The archetypal geometric objects were perfectly drawn in lines with zero width. And it had to be possible to construct a proof using only a small number of items – something like a pair of compasses, a straight edge and a perfect pencil to produce those zero-width lines.

Many challenges in geometric construction were overcome and were included in Euclid's remarkable books on geometry, which started with a small number of 'givens' or 'axioms' and built a series of logical proofs. So robust was this system, because it was built on an abstract absolute rather than the real world, that Euclid's books were still being used to teach geometry in the 20th century.

However, there were a number of apparently simple constructions and proofs that evaded the early geometers, so a lot of effort was put into attempting to complete them. One example was trisecting an angle – dividing it into three equal parts – and another was squaring the circle, which involved constructing a square with the same area as a given circle. Like the trisection this would eventually be proved impossible. We can see that easily now because we know that the area of the circle is dependent on pi, a transcendental number – one that never settles down to a repeating pattern, which makes it impossible for a simple construction technique to reproduce the area exactly.

*Further reading: **Are Numbers Real?***

Elementary, my dear Aristotle

Name Aristotle's five elements.

Answer overleaf ➔

While you're thinking ...

Aristotle lived between 384 and 322 BC and was a pupil of Plato.

The idea of *four* elements dates back to Empedocles,
born just over 100 years before Aristotle.

Aristotle is one of the many individuals to have been called
'the first scientist', though his philosophical methods were
very different from those of a modern scientist.

Earth, air, fire, water and quintessence

All five required to get a point.

The four, more familiar, ancient elements date back further in Ancient Greek history and were based on reasonably logical, if incorrect, thinking. If you imagine, for instance, burning a piece of green wood, it gives off fire, watery sap oozes out of it, hot air and steam come off it, and you end up with ash. It seems to break down to those four elemental forms.

Much Ancient Greek physics was based on the idea that earth and water had gravity; their natural condition was to be as close to the centre of the universe as possible (i.e. close to the centre of the Earth), while air and fire had levity, with a tendency to move away from the centre of the universe.

Aristotle added the fifth element because he believed that everything below the Moon's orbit was transient, but everything above it was eternal and unchanging. He felt that this perfect, immutable outer region needed its own special element that was not subject to levity or gravity, and it was simply named the 'fifth element' – the quintessence.

Because everything above the Moon's orbit was expected to be unchanging (apart from the cyclical rotation of the heavenly spheres), Aristotle's cosmology struggled with the likes of comets and new stars (what would become known as novas or supernovas). To fit his worldview these phenomena had to be sublunary – below the orbit of the Moon – which led to some mental contortions to explain how this could be possible.

Further reading: **Reality's Frame**

A modest mind

Who, when asked if it was true that only three people in the world understood general relativity (including himself and Einstein), is said to have answered 'Who is the third?'

Answer overleaf →

While you're thinking ...

Einstein published his definitive paper on general relativity
(more correctly his general theory of relativity) in 1915.

While it's possible to understand Einstein's earlier special theory of
relativity with high school mathematics, even the great man himself had
to get help with the maths that would be required for general relativity.

The person who made the remark was, at the time,
one of the most famous physicists in the world.

Arthur Eddington

There are a number of variants on the Eddington story. In the simple form used in the question, the questioner was said to be a newspaper reporter, while in a more sophisticated version, another physicist, Ludwik Silberstein, was said to have commented to Eddington that Eddington was one of the three who understood the theory. Silberstein, who clearly had himself and Einstein in mind for the other two, mistook Eddington's delay in replying for modesty and told Eddington not be so shy, at which point Eddington is said to have replied cuttingly, 'Oh, no. I was wondering who the third might be.'

There was no doubt that Eddington did understand the theory (though in reality there were plenty of others who did, including German mathematician David Hilbert, who, seeing an early version of Einstein's ideas, nearly beat Einstein to the publication of a final version). Eddington also made considerable contributions to cosmology. But part of the reason he was so well known was because he was the Brian Cox of his day – he was an excellent science populariser.

It was also Eddington who helped cement Einstein's victory by arranging an expedition to measure the light-bending effect predicted by general relativity during the solar eclipse of 1919. It has since been suggested that Eddington made sure the results supported Einstein's theory, as his equipment and the conditions made it very difficult to come up with a definitive answer, but as far as the world was concerned, Eddington's observations turned Einstein from a reasonably well-known theorist into a scientific superstar.

Further reading: **Reality's Frame**

QUIZ 1
ROUND 3:
TECHNOLOGY

Jet set

Which jet engines were developed at the US Jet Propulsion Laboratory?

Answer overleaf ➜

While you're thinking ...

The Jet Propulsion Laboratory (JPL) is located in La Cañada,
California, but this city did not exist when the JPL was first
set up, so it's usually described as being in Pasadena.

Although ideas for jets had been around for some time, the invention
is usually ascribed to the British engineer Frank Whittle, who applied
for a patent in 1930 and had a first working engine in 1937.

The first aircraft to fly with a jet engine was a
German Heinkel He 178 in August 1939.

No jet engines were developed there

Somewhat surprisingly, the JPL in Pasadena has never been involved in designing or building jet engines. The laboratory was, instead, the main US development centre for rocketry and has more recently built satellites. According to JPL legend, when rockets were first on the table in the 1930s, they were considered a topic of science fiction rather than a serious governmental development, and so the work was given the less controversial cover of jets. (The same goes for the 'Jet' in JATO, or 'Jet Assisted Take-Off' used in aircraft carriers: it's a rocket.)

This scornful attitude is reflected in the treatment received by Robert H. Goddard, now feted as the USA's rocket pioneer, when he published a paper based on his experiments in 1920. The press picked up on it, treating Goddard's claim that a rocket from Earth could reach the Moon as a huge joke. They called Goddard the 'moon man' – intending this as an insult. The worst of the attacks came from *The New York Times*, which published an editorial commenting 'That Professor Goddard, with his "chair" in Clark College and the countenancing of the Smithsonian Institution, does not know the relation of action to reaction, and of the need to have something better than a vacuum against which to react – to say that would be absurd. Of course he only seems to lack the knowledge ladled out daily in high schools.'

The New York Times had misunderstood how rockets work and would only correct their error while Apollo 11 was in flight, saying that further investigation and experimentation had confirmed the findings of Isaac Newton. It's a shame that this wasn't obvious to the newspaper's writers 49 years earlier.

Further reading: **Final Frontier**

Glow-in-the-dark gadgets

What item that should be in every house depends on a radioactive source to work?

Answer overleaf ➜

While you're thinking ...

There are plenty of mildly radioactive objects in any home.
Human beings are slightly radioactive, for example. But this
is an object where a radioactive source is actively used.

For much of the 20th century, a popular US crockery line known as
Fiestaware was coated in uranium dioxide glaze to produce impressive
orange-red colours (and set off any nearby Geiger counters).

The natural background level of radioactivity varies considerably
from place to place. Edinburgh, for instance, with its ubiquitous
granite, has higher levels than London, while Cornwall is
even higher, at around three times the London level.

The smoke detector

'Radiation' is a very general term for something that passes from place to place in the form of a stream of quantum particles. Light, for instance, is electromagnetic radiation. So it is common to refer to the kind of radiation emitted by radioactive sources as 'ionising radiation'. This is because when the radiation hits matter, it tends to ionise it, knocking electrons off atoms to form charged particles. When this happens in living tissue, it can cause damage leading to cancer from long-term medium levels of exposure and radiation sickness from high levels.

However, this doesn't mean that ionisation is always a bad thing – and in a smoke detector it is used to save lives. A typical detector contains a tiny piece of the radioactive element americium-241. This gradually decays, giving off ionising radiation in the form of alpha particles, charged helium nuclei, which ionise the air in a small chamber inside the detector. Ionised air conducts electricity (this is how lightning flows), and a tiny current is maintained across the ionised air, but when smoke gets into the detector, it absorbs the alpha particles before they can ionise the air, the current stops and the detector sounds.

In principle you could use the americium from smoke detectors to make an atomic bomb: get enough of the material together in a lump and it would explode. However, there is such a tiny amount in any one detector you would need to pull apart 180 billion of them to retrieve the element to construct your device, so the security services are not unduly worried.

Further reading: **Armageddon Science**

Seeing from afar

Who invented the telescope?

Answer overleaf ➔

While you're thinking ...

The word 'telescope' is derived from the Greek words for 'far looker' and arrived in the English language via an alliance of Latin and Italian.

Most early telescopes used lenses to focus light, but within a couple of generations, the best astronomical telescopes used curved mirrors, which focused all colours equally and could be a lot smaller for the same magnification compared to a lens-based refracting telescope.

We continue to produce larger and larger telescopes. At the time of writing, the biggest planned is the E-ELT that, which will have a 39-metre main mirror (nearly eight times the size of the 200-inch Mount Palomar telescope which was the world's largest for many years).

We don't know who invented the telescope

What is certain is that it was not Galileo who invented the telescope – take a point away if you thought (like TV show QI) that Galileo was first. We know this, as Galileo built his first telescope in response to the arrival in Italy of a Dutch optician who had a telescope to demonstrate in Venice – Galileo managed to get there first with his newly built instrument.

Another wrong answer (but closer, so no points lost) was Hans Lippershey, a Dutch optician who tried to patent the telescope at the start of the 17th century, but failed because at least two other spectacle makers claimed to have already made such devices.

It is by no means certain, but one definite earlier possibility (for which you earn half a point) was the Elizabethan father and son pairing of Leonard and Thomas Digges. Thomas appears to have built an instrument based on his father's ideas towards the end of the 16th century, based on a lens–mirror combination. The British court asked William Bourne, an expert on military technology, to check the invention out. He described the telescope and some of its limitations (it had a narrow field of view), suggesting that there was an actual device for him to examine.

However, some kind of telescope may have existed much earlier. The 13th-century friar and proto-scientist Roger Bacon wrote in 1267 that 'we can so shape transparent bodies, and arrange them in such a way with respect to our sight and objects of vision, that the rays will be reflected and bent in any direction we desire, and under any angle we wish, we may see the object near or at a distance […] So we might also cause the Sun, Moon and stars in appearance to descend here below.'

*Further reading: **Light Years***

WHAT COLOUR IS THE SUN?

Monkish business

What is an 11th-century Benedictine monk called Eilmer supposed to have done after jumping off the tower of Malmesbury Abbey?

Answer overleaf ➜

While you're thinking ...

We know about Eilmer because he was written about by another monk who lived later in the same abbey, the historian William of Malmesbury.

The remains of the Benedictine monastery still stand in the Wiltshire town. Although, like many abbeys, it has become a ruin since the dissolution of the monasteries, part of the Abbey church remains in use as Malmesbury's parish church.

The abbey tower collapsed in a storm around 1500, taking much of the existing church with it – so we don't know exactly what Eilmer's perch looked like.

The monk Eilmar is supposed to have flown

Eilmer is often cited as the man behind one of the first human flights, and though nothing from this period can be considered to have perfect historical accuracy, William of Malmesbury's account was written sufficiently soon after the event to be likely to have a reasonable level of accuracy.

If Eilmer did achieve the success that was claimed, of flying 'more than a furlong' (so over 200 metres), it is because unlike many early attempts at flying machines, he did not try to imitate a bird with flapping wings, but rather the effortless soaring of the wind-borne raptor. Inspired by the Greek myth of Daedalus and Icarus, Eilmer is said to have attached wings to his arms and possibly his legs and launched himself off the tower.

According to the account, the flight did not end well, as Eilmer is said to have broken both his legs as he hit the ground, and to have been lame for the rest of his life. He apparently blamed his bad landing on forgetting to equip himself with a tail like a bird, and certainly the lack of a tail would have made his flight very unstable.

It wasn't until the early 19th century that the engineer George Cayley succeeded in flying a model glider, and several decades later when the first manned flight using one of his designs took place.

Further reading: **Inflight Science**

Cyborg cockroach

What can you do with a Roboroach?

Answer overleaf →

While you're thinking ...

Generally, the term robot is used to describe a totally mechanical
organism, while an android is an artificial organic lifeform,
though the terms were originally used the other way round.

The term cyborg, a contraction of 'cybernetic organism',
seems to have been first used around 1960.

Despite their reputation, cockroaches are not able
to survive high levels of nuclear radiation.

Remotely control the movement of a living insect

Although the glamorous (and sometimes more horrific) face of cyborg technology is the combination of human and technology, much of the work that has been done for real has been with insects, which have the advantages of being cheap, readily available, relatively simple as organisms and rarely given too much concern by animal rights organisations.

The military, for example, have shown interest in the ability to control flying insects by inserting technology into their nervous system. If these insects can then carry, for example, a radio microphone and camera, they become in effect tiny, highly manoeuvrable drones. From the operator's viewpoint, they are simple to fly, as controls tend to handle directions, but the insect's internal systems manage all the difficulties of stable flight and avoiding obstacles. To date, this kind of insect cyborg has tended to be based on quite large beetles, to be able to carry the kit, though the ideal would be to achieve sufficient miniaturisation that a fly could be used.

The rather misnamed Roboroach is a commercial alternative, where it's possible to buy a kit to wire up a (non-flying) cockroach and steer it like a remote-controlled car. The kit comes in a 'supply your own cockroach' form, though insects can be provided for those without a handy source. Electrodes are inserted in the base of the insect's antennae, linking to a tiny radio backpack, which enables the cockroach to be controlled by a smartphone. The manufacturers claim that this is an educational tool to learn about both cybernetics and neuroscience.

Further reading: **Ten Billion Tomorrows**

Swinging time

Who invented the practical pendulum clock?

Answer overleaf ➜

While you're thinking ...

A pendulum was used in the 2,000-year-old earthquake detector of Zhang Heng, where the movement of the pendulum as the earth shook was used to activate a lever and deliver a ball into the mouth of a toad, showing the direction of the disturbance.

Galileo, famously, is said to have studied the motion of a chandelier in Pisa cathedral, noticing how the time of its swing was not connected to how far it moved.

The observation that the timing of the pendulum is only dependent on its length, not the size of the swing, only applies for small amplitudes. If the pendulum swings too far, this no longer applies.

Christiaan Huygens invented the practical pendulum clock

Although Galileo did indeed observe the way that the timing of a pendulum's swing was independent of the size of that swing (and the weight of the bob), and realised that this would make it a good regulator for a clock, he never built one. He did describe a design to his son Vincenzo, who began work on it but died before completing the clock.

It fell to Christiaan Huygens to produce the first working pendulum clock in 1656. The Dutch scientist, like many of his contemporaries, was a polymath. He worked on everything from the laws of motion and a wave theory of light to astronomy, discovering Saturn's moon Titan. But in practical terms, his most impressive contribution was probably the pendulum clock. This was a huge leap forward for mechanical clocks, which previously had been notoriously inaccurate. With a pendulum as a regulator, the accuracy of clocks improved nearly 100-fold.

Huygens published a detailed theory on the workings of the pendulum clock. He was aware of the limitation that the timing only stayed constant for small swings, which proved a problem as the early clock mechanisms tended to swing the pendulum around four times as far as could safely be done to keep a constant time. As a result, much of the time spent on developing clockwork mechanisms for the next few decades was devoted to improving mechanisms called escapements which would work with a much smaller pendulum amplitude.

*Further reading: **The History of Clocks and Watches***

Check mech

How did the Mechanical Turk chess automaton from the 1760s manage to beat human chess players?

Answer overleaf ➜

While you're thinking ...

Arguably the oldest automaton in mythology was the artificial man Talos, made by the Greek god Hephaestus to protect Europa, described in detail in the story of Jason and the Argonauts, though first mentioned more than a century earlier, around 400 BC.

Leonardo da Vinci made some remarkable automata including a mechanical knight and a lion that walked, wagged its tail and stopped to open a panel showing the royal fleur-de-lis.

In 1779 C.G. Kratzenstein produced an automaton speaking machine, though it could only handle the vowels. A little more than ten years later, Wolfgang von Kempelen, who made the Mechanical Turk, had produced a hand-controlled speaking device that could produce recognisable words.

A human chess player sat inside and operated it

The Mechanical Turk was a wonder of the 18th century which would outlive its maker and go on to be exhibited across the world for decades. It consisted of an elegant wooden chest with a chess board on top, behind which sat a mechanical figure, dressed in a turban. The Turk was clearly an automaton – it was possible to see some of the mechanism behind doors in the chest – and yet it was able to play a very effective game of chess, often beating human players, including Napoleon I of France.

Built in the 1770s, the Turk went on to amaze audiences, first in Europe and then in the USA, until it was destroyed in a fire at the Chinese Museum in Philadelphia. There is no doubt that the Turk was a clever device, able to pick up and move chess pieces; but despite appearing to be an early predecessor of the Deep Blue chess computer that beat world champion Garry Kasparov, the Turk had no intelligence, relying instead on human thought to make its moves. A chess player sat inside the complex cabinet, which included, as well as misleading mechanisms, a moving seat, which the operator could shift from side to side when various parts of the cabinet were opened to create the illusion that there was nobody inside. It was also equipped with a magnetic repeater, so the operator could see progress on the board above.

By using a control that reproduced the motion of the Turk's arm on a dummy board, the operator could make the Turk carry out all the required moves. The Turk was disassembled and rebuilt several times during its career, first because its maker, Wolfgang von Kempelen, became bored with displaying it, and later when it changed hands after von Kempelen's death.

*Further reading: **The Mechanical Turk***

WHAT COLOUR IS THE SUN?

Home bytes

Which early personal computer maker had a PET and a VIC20?

Answer overleaf ➜

While you're thinking ...

The concept of a personal computer only became possible
outside of a handful of extremely expensive devices with
the development of the microprocessor, putting much
of the electronic circuitry needed on a single chip.

The first microprocessors were developed in 1971 by three
different US manufacturers, including Intel's 4004 chip.

The Apple I computer, one of the first personal
computers available to hobbyists (as a circuit board),
was introduced in 1976, the year before the PET.

Commodore

While in the USA companies like Apple and Tandy dominated, the home computer market in the UK was mostly owned by domestic companies like Sinclair and Acorn (makers of the BBC Micro that was the UK school standard in the 1980s). But one company became highly successful on both sides of the Atlantic: Commodore. Started in Toronto by Jack Tramiel in 1954 as an office equipment company, before moving into watches and calculators and then computers, Commodore is probably the biggest name from 1980s computing that is pretty much forgotten today.

The Commodore PET 2001 was one of the very first home computers, launched in 1977 with a monitor and keyboard built into a single metal casing. But the PET was expensive, and it was the VIC20 (named after a chip, the Video Interface Chip) that would first give Commodore mass-market success. This was followed up with the wildly successful Commodore 64 (a reference to the 'massive' 64 kilobytes of memory), which, with the Sinclair Spectrum, became one of the two British home standards.

Arguably Commodore's computer output peaked with the Amiga. This was a home computer ahead of its time, with excellent graphics that set the standard for the games of the period, and which even had a basic graphical user interface as standard. But by this time, IBM-compatible PCs were beginning to dominate, and though it would be years before they caught up with the Amiga's graphic and sound capabilities, the sheer size of their market and their flexibility meant that they pulled the rug from under the Amiga and Commodore. The company ceased to exist in 1994.

Further reading: **Electronic Dreams**

QUIZ 1
ROUND 4:
MATHEMATICS

Taking sides

What is a chiliagon?

Answer overleaf →

While you're thinking ...

Many mathematical terms ending in 'gon' are two-dimensional
shapes, for example a pentagon. We take pentagon from
the Greek for 'five angles', via the Latin *pentagonus*.

A chilindre was a medieval portable sundial. Chaucer wrote: 'And lat us
dyne as soone as that ye may ffor by my chilyndre it is pryme of day.'

In working out the size of the universe, Archimedes declared,
'the diameter of the Sun is greater than the side of a
chiliagon, inscribed in the greatest circle of the universe'.

A chiliagon is a regular two-dimensional shape with 1,000 sides

Just as a pentagon has five sides, the (significantly rarer) chiliagon has 1,000 sides, though it is not a shape that is often used in modern mathematics.

It's just a coincidence that Archimedes linked a chiliagon to the Sun and a chilindre was the name for a portable sundial. The name of the timepiece was just a variant on the medieval Latin word *chilindrus*, which was a variant of *cylindrus* – it simply meant that the device was cylindrical.

Archimedes, however, was making reference to the 1,000-sided shape. In a short book called *The Sand Reckoner*, he was attempting to work out how many grains of sand it would take to fill the universe. (This wasn't a practical suggestion; he was using it as an example of how to extend the very limited Greek number system.) Before Archimedes could do the calculation, he had to work out how big the universe was. This involved various assumptions and approximations.

In his book, Archimedes explains that he makes the chiliagon assumption because the astronomer Aristarchus had found that the diameter of the Sun appeared to be around $\frac{1}{720}$ of the circle of the zodiac. Archimedes then goes on to describe an experiment that he undertook to attempt to confirm this. As a result of his calculation, including the role of the chiliagon, Archimedes calculated that the universe (he meant what we would now call the solar system) was what we now know to be around the size of the orbit of Saturn – not bad given the limited information available to him.

*Further reading: **A Brief History of Infinity***

Mental challenge

Without using pen and paper or a calculator, what is 1 + 2 + 3 ... + 100?

Answer overleaf ➜

While you're thinking ...

The '...' mark is called an ellipsis, and indicates that something has been omitted without losing the meaning, or 'and so on'.

In mathematics, the ellipsis mark is used slightly differently to mean 'continue doing the same until indicated otherwise'. So 1 + ½ + ¼ + ⅛ ... indicates that the series goes on for ever. However, it is also possible to have such a series with a limit; in the case of our question, we require the sum of every whole number from 1 to 100 inclusive.

The word ellipsis has the same root as 'ellipse', from the Greek for 'coming short'. This is because an ellipse is a conic section (the result of making a cut through a cone) that 'comes short' of the base – it doesn't cut into the base of the cone.

Five thousand and fifty

It would be possible to work this out by sequentially adding together the numbers from 1 to 100, but that would take quite a while, and would be difficult to do in your head. However, the great German mathematician Carl Friedrich Gauss is said to have amazed his primary school teacher when set this challenge by coming up with an answer in seconds (as is often the case, there is some doubt whether this legendary event took place). And it's not necessary to be a mathematical genius to do this trick – as long as you can come up with the method Gauss is presumed to have used, anyone with basic mental arithmetic skills can do it too.

The trick to calculating the sum is slightly more obvious if we write out a symmetrical set of entries at both ends of the series. If we work on 1 + 2 + 3 ... + 98 + 99 + 100 it becomes slightly more obvious that the first and the last entry in the series (1 + 100) add up to 101. So do the second and the penultimate (2 + 99). And the next pair (3 + 98), and so it goes through all the numbers. There are 50 pairs of this kind. So the total is just 50 × 101, or 5,050.

There are plenty more of these mental arithmetic workarounds that make it easy to work out apparently complex calculations. For example, you can often turn a difficult subtraction into an easy addition by over-subtracting. So 8,262 – 669 is much easier to do in your head by rounding up the 669 to 700, making the subtraction the easy 8,262 – 700 (7,562), then adding 31 (700 – 669) to make 7,593. Many of the tricks, however, are harder to commit to memory.

*Further reading: **The Magic of Maths***

Calculating conundrum

Why is calculus called 'calculus'?

Answer overleaf ➜

While you're thinking ...

Calculus is a mathematical method that incorporates two main (and opposite) roles: the first calculates the way one variable changes with another (for instance the acceleration when something's velocity is changing with time), known as differential calculus. The second involves summing up a series of small segments and reducing the segment size to practically nothing, known as integral calculus.

The development of calculus was subject to a huge priority dispute between Isaac Newton and Gottfried von Leibniz, who seem to have devised the approach independently. Newton referred to calculus by the clumsy term 'the method of fluxions', while Leibniz used the name (and symbols) we use today.

The Royal Society in London set up a commission to discover who came up with calculus first and decided it was Newton. The commission's report was written by the president of the Royal Society, but some might doubt its impartiality as the president at the time was Isaac Newton.

Calculus is named after *calculi*, the counting stones used in the Roman equivalent of an abacus

In Latin, *calculus* just means a small stone (the diminutive of the word *calix* from which we get terms like calcify). The Romans used counting boards which were wooden boards (or just a flattened area of earth) marked out into columns. Stones were placed into the columns to perform calculations, much as the beads are moved on different sections of an abacus.

Originally, calculus was just used, by extension of the idea of a counting stone, as a general word for a method of working something out mathematically. It wasn't specifically applied to what we'd now call calculus at this stage, and it's in this sense that Leibniz is likely to have used it when applying it to both the differential and integral forms he developed. However, within a few decades, the overall approach developed by Newton and Leibniz was known by the grand term 'the calculus', and by the mid-20th century the article was frequently dropped. Now, anyone referring to calculus would be assumed to be dealing with Leibniz's formulation.

Further reading: **A Brief History of Infinity**

It's all Greek to me

What language are the terms algebra and algorithm derived from?

Answer overleaf ➔

While you're thinking ...

As well as the more familiar mathematical use, algebra also used to be a term for treating a bone that was broken or dislocated.

Algebra is a type of mathematics where varying quantities are often represented by letters in formulae where it is possible to work out the value of one or more of these varying quantities given the values of others. To some students, algebra is baffling; to others it represents an entertaining mathematical puzzle. Either way, it is a powerful technique for analysis.

Algorithms, while originally purely mathematical, are now highly important in computing as they describe a procedure or set of rules that is used to carry out a mathematical operation and so can make up a part of a computer program.

The terms algebra and algorithm are derived from Arabic

The 'al' prefix is a clue. Algebra certainly comes from the Arabic, and while there has been some suggestion that algorithm has a Greek origin, the argument, as we shall discover, is rather weak.

In the 12th century, the medieval scholar Robert of Chester translated part of an Arabic mathematical text called *al-kitāb al-muḵtaṣar fī ḥisāb al-jabr wal-muqābala*, which approximately means 'the concise book on calculation by restoration and compensation'. The *al-jabr* part was extracted and used for mathematical calculations, initially those involving quadratic equations (like $2x^2 + 3x + 4 = 0$), but then extended to include more general algebraic equations.

When it comes to algorithms, the slightly dubious link to Greek, given as the word's origin in the *Oxford English Dictionary*, is that it has the same origin as arithmetic, which was originally *arismetica*. *Algorismus* is the Latin form from which algorithm is derived. So it's not very close. A more plausible origin is in the name of the author of the translated Arabic text referred to above, who was called al-Khwārizmī (or more completely Abū Ja'far Muḥammad ibn Mūsā al-Khwārizmī). This was Latinised as Algoritmi.

*Further reading: **Are Numbers Real?***

Number love

Why are 220 and 284 called amicable numbers?

Answer overleaf →

While you're thinking ...

A whole range of special numbers are given names, a practice dating back to the Ancient Greeks (though not all the terms go back that far).

A perfect number, for example, is a number that is the sum of all the positive numbers (other than itself) that divide into it. So, for instance, 6 is the smallest perfect number as 1 + 2 + 3 = 6. It's followed by 28 and 496, but the sequence rapidly grows, so that by the time we reach the ninth perfect number, it has 37 digits.

Some mathematicians argue that all numbers are special. Otherwise, if you imagine the smallest number that isn't special, then it becomes special because of being in this position. And so on.

Because the factors of each number add up to the other

The Pythagoreans were fascinated by whole numbers and their relationships, ascribing them with special properties. Ten, for instance, was considered to represent perfection, as opposed to the perfect numbers mentioned on the previous page. This is because, firstly, it is the sum of the first four numbers, and also ten items can be arranged to form a perfect triangle.

It might seem a little odd that 220 and 284 appealed to the Pythagoreans, but the factors of 220 (the numbers that it can be divided by) are 1, 2, 4, 5, 10, 11, 20, 22, 44, 55 and 110. Add these up and you get 284. Similarly, the factors of 284 are 1, 2, 4, 71 and 142, which added together make 220. Although there is no significance to this, the reciprocity appealed to the number-oriented Ancient Greeks and so these became known as the 'amicable numbers'.

Such is the appeal of this pairing that it is possible to buy a pair of pendants in the form of a split heart, with one number on the first part, and the other on the second – the ideal example of a nerdish love token.

*Further reading: **Are Numbers Real?***

Symbolic significance

What is a lemniscate?

Answer overleaf →

While you're thinking ...

The word 'lemniscate' comes from the Latin
lemniscatus meaning 'adorned with ribbons'.

Mathematical terms can be quite misleading. A transcendental number,
for instance, does not contemplate its navel, but is a number such
as pi that is impossible to calculate exactly with a finite equation.

The lemniscate first appeared in 1655.

Lemniscate is the name of the potential infinity symbol, ∞

It is relatively unusual that we can date the origin of a mathematical symbol exactly, but the first use of ∞ was in *A Tract on Conic Sections*, written in 1655 by English mathematician John Wallis. Infinity was not a new concept. The Ancient Greeks had discussed it at length, and Galileo had made some early mathematical observations on its nature. But until around the time that Wallis came up with the symbol, infinity had limited use. However, the introduction of calculus by Leibniz and Newton would make infinity an important limit in some mathematical operations.

You don't lose a point if you just say that the lemniscate is the symbol for infinity, but it's more accurate to say that it represents potential infinity – the never-reached limit used in infinite sums and in calculus. The true infinity that is the size of the set of the integers (whole numbers) has its own symbol: $\aleph_0$ or aleph null.

Lemniscate is also used as a generic term for geometrical shapes that resemble the number 8 and for a class of algebraic functions which are a subset of the elliptic functions. We aren't sure, however, why Wallis chose this shape. Some suggest it's a loop that goes on for ever, and others that it resembles a symbol used by the Romans for 1,000 as an alternative to M. But Wallis merely said that he would let ∞ represent infinity.

Further reading: **A Brief History of Infinity**

How long is a piece of string?

Which branch of maths did 'rope stretchers' use in ancient Egypt?

Answer overleaf ➜

While you're thinking ...

Ropes have been made since prehistoric times. It seems likely, from impressions in clay that was subsequently hardened in a fire, that woven fibres were made into cord as long as 28,000 years ago.

The Egyptians were probably the first to go beyond basic twisted fibres, developing tools specifically for rope making.

Egyptian ropes were most frequently made from the fibres of plentiful water reeds.

Geometry

Ancient Egyptian 'rope stretchers' were surveyors. The term refers to the way that surveyors marked out an area of ground using ropes stretched tight between posts in order to provide straight-line sides which did not droop more than could be helped. This kind of work was employed both in constructing buildings and calculating the position of destroyed boundaries after the regular Nile floods.

When measuring out their spaces for foundations, Ancient Egyptian surveyors and masons were among the earliest exponents of geometry. Long before Pythagoras, for instance, they were using the 3:4:5 relationship of Pythagoras' theorem to calculate the size of sides of right-angled triangles.

One of the oldest known documents containing examples of geometry problems is the Ancient Egyptian Rhind Papyrus, which is around 4,000 years old. As well as showing how to work out simple areas, it provides methods for calculating the volume of granaries on both rectangular and circular foundations, and for calculating the *seked*, a measure of the slope of a pyramid.

*Further reading: **Are Numbers Real?***

The Twilight Zone

What is the Monster group?

Answer overleaf ➔

While you're thinking ...

The word 'monster' is most frequently used for a horrible imaginary
creature nowadays, but it has also been used since medieval
times to mean something extraordinary and marvellous.

Groups were introduced into mathematics in the 19th century. They are
special sets of mathematical elements that obey specific rules, including
that there must be an operation that combines any two of the elements
to make a third element. The integers, for example, form a group.

Probably the best-known application of groups is in
the 'symmetry groups' that are used in physics to
structure particles and their interactions.

The nickname for the mathematical group reflecting the different ways something could be rotated if you had 196,883-dimensional space available.

You might argue that we don't have 196,883-dimensional space available, but mathematicians don't worry about that. Although we are familiar with the three dimensions of space (or four of space-time), mathematically speaking it is possible to go on adding dimensions as far as you like. These aren't directions in which to move; rather they are measures which are independent of each other but which can sometimes be linked in such a way that one influences another.

When mathematicians work with mega-multidimensional mathematical objects, they may just be playing for the sake of it. A lot of mathematics starts out that way. But such multi-dimensional work can also be very useful. For example, in quantum physics calculations on a system are undertaken by treating each of its component particle's properties in a separate, imaginary dimension.

It's pretty much impossible to explain what the Monster group is without a lot of mathematical background. It's the largest sporadic simple group, where a simple group is one that doesn't have any subgroups obeying a particular rule, and the sporadic simple groups are the simple groups which don't fit into any of the eighteen identified families of such groups. Suffice it to say that only a mathematician could love it.

Further reading: **Things to Make and Do in the Fourth Dimension**

QUIZ 1
ROUND 5: PHYSICS

Jolly genius

What was it that Einstein described as his 'happiest thought'?

Answer overleaf →

While you're thinking ...

Like many physicists, Einstein was fond of music and played the violin to the best amateur concert standard.

Einstein famously remarked that 'When a man sits with a pretty girl for an hour, it seems like a minute. But let him sit on a hot stove for a minute and it's longer than any hour. That's relativity.' He claimed this was a result of an experiment, with the help of movie star Paulette Goddard, whom he met through his friend Charlie Chaplin.

Einstein, probably more than any other physicist, was fond of making use of 'thought experiments' – hypothetical experiments, played out in the mind, to test out a theory or produce a new one.

The equivalence principle

Also score a point for any variant of 'if you fall freely, you don't feel your weight'. In 1907, by his own account while sitting in his chair in the Patent Office in Bern, Switzerland, before his first academic position, Einstein had this 'happiest thought'. What he realised was that the acceleration produced by falling freely cancelled out the pull of gravity. This is why astronauts in the International Space Station float around – the gravitational pull there is still nearly as strong as on Earth's surface, but in orbit they are falling towards Earth (it's only the station's sideways motion that stops them from crashing). From here, Einstein made the leap to suggest that acceleration and gravity could not only cancel each other, but were equivalent and indistinguishable.

Imagine, for instance, being in a spaceship parked upright, with its engines beneath it, on Earth. You would feel a pull of gravity towards the floor of the craft. Now imagine the same ship in space with its engines running, giving it the same acceleration as Earth's gravity. Again, you would feel a pull towards the floor of the craft, just as you are pushed into your seat when a plane accelerates on the runway. Einstein suggested that the two were indistinguishable.

However, when a ship is under powerful acceleration, a beam of light crossing the ship will no longer appear to travel in a straight line, but will bend as the ship accelerates away. So gravity should also bend a beam of light. Einstein extended this thought to realise that gravity warps space and time, leading ten years later to his development of the general theory of relativity.

*Further reading: **Reality's Frame***

Law maker

Who first set out Newton's first law of motion?

Answer overleaf ➔

While you're thinking ...

Newton's three laws of motion are described
in his masterwork, *The Principia*.

He stated the first law as 'Every body perseveres in its state of
being at rest or moving uniformly straight forward, except insofar
as it is compelled to change its state by forces impressed.'

In our everyday world, the first law is counter-intuitive as things tend
to stop moving because of friction and air resistance unless something
keeps pushing them. But these resistive forces are just some of
Newton's 'forces impressed'. Without such forces – in empty space,
for example – things that are moving do indeed keep doing so.

Galileo (or Aristotle)

Give yourself a point for either, but take one off for Newton. Galileo, well before Newton was born, came to the same conclusion. He worked this out from his experiments with inclined planes, where he rolled balls along very smooth slopes. He discovered, not entirely surprisingly, that balls got faster rolling downhill and got slower rolling uphill. The clever part was his deduction that on the level, a ball should neither accelerate nor decelerate, but continue at constant speed.

Aristotle was a more interesting, indeed a perverse, case. The Ancient Greek philosopher was convinced that a vacuum, or more accurate a void – space with absolutely nothing in it – could not exist. And the argument he used was essentially to say that, if a void did exist, Newton's first law would apply. And since we never see it applying, then there is no void.

The argument Aristotle used went like this: in such a void, 'no one could say why something moved will come to rest somewhere; why should it do so here rather than there? Hence it will either remain at rest or must move on to infinity unless something stronger hinders it.' In essence, this was a perfect statement of Newton's first law, made 2,000 years before Newton was born. The only problem was that Aristotle was using it as absurdity, rather than something he thought would truly happen.

Further reading: **Reality's Frame**

Emmy award

In which area of physics did Emmy Noether's theorem first prove significant?

Answer overleaf ➔

While you're thinking ...

Emmy Noether was arguably the 20th century's foremost
female mathematician, lecturing at the University
of Göttingen in the 1920s and early 1930s.

Noether was initially known by her first name, Amalie,
but switched to her middle name as a child.

In her undergraduate studies, Noether was one of only two
female students attending the University of Erlangen.

The conservation laws

Developed in 1915 and published in 1918, Noether's theorem made it clear just how important symmetry is to physics. It linked the conservation laws to symmetry considerations. Conservation laws establish that quantities like the amount of energy, momentum and angular momentum in a closed system are conserved. That is, these quantities stay the same unless something enters or leaves the system. And each conservation law follows on directly from the principle that the system in question has a certain sort of symmetry.

Symmetry in this context is not just the familiar mirror symmetry, but that something should appear the same after undergoing some kind of change. For example, there is rotational symmetry, where something appears the same after turning it around, or translational symmetry, where it remains the same (typically if present in an infinite repeated pattern) when the whole system is shifted sideways by a certain amount.

Noether showed that each conservation law emerged from a particular symmetry. So, for instance, conservation of energy is a result of symmetry in time – where a shift in time does not change the way a process takes place in a system. Later, physicists would take deductions from symmetry to a whole new level, using symmetry considerations to establish many aspects of the standard model of particle physics used today.

*Further reading: **Reality's Frame***

Play your quarks right

What name, derived from playing cards, might quarks have had?

Answer overleaf ➔

While you're thinking ...

The name 'quark' for the fundamental particles that
make up more familiar particles like neutrons and protons
was devised by physicist Murray Gell-Mann.

It is often said that Gell-Mann took the name from James Joyce's
novel *Finnegans Wake*. He didn't. It just came to him, as a sound
he liked. That sound was *kwork*, so strictly speaking, 'quark' should
be pronounced 'kwork'. However, when Gell-Mann saw Joyce's
meaningless phrase 'three quarks for Muster Mark', he chose
to spell the word this way, as quarks often come in threes.

The word 'quark' was used for a soft German cottage cheese
well before either Joyce or Gell-Mann employed it.

Quarks might have been called aces

The US physicist George Zweig, who was working at the CERN laboratory near Geneva, Switzerland, came up with a very similar idea to Murray Gell-Mann's quarks, but Zweig called them aces, referring to the playing cards, as he believed at the time that there were four such particles.

Both Zweig and Gell-Mann were trying to find some underlying explanation for the burgeoning particle zoo that was being discovered in laboratories. They argued that making use of an eight-fold symmetry enabled them to build a model of particles like protons and neutrons constructed from triplets of varying simpler particles, while a different family of particles, the mesons, contained pairs of these sub-particles.

Both physicists derived their models in 1964 and they would begin to be experimentally confirmed a few years later. Although the extreme nature of the strong force that holds them together means that we don't see 'naked' quarks, their existence is very strongly implied by many experiments and there remains no significant doubt of their existence. The adoption of quark rather than ace seems largely down to Gell-Mann's presence in the USA, where at the time he dominated the field. Had the theory been developed now, with CERN's pre-eminence, it's entirely likely that we would be referring to aces.

*Further reading: **Schrödinger's Kittens***

Particle flavours

How many flavours of quark can you name? (A point for each)

Answer overleaf →

While you're thinking ...

There are a total of six flavours of quark, each with an antiquark – we're only looking for the ordinary quarks, so there are no extra points for naming antiquark flavours.

Physicists have an unfortunate habit of using terms that bear no resemblance to the way those words are normally used. Flavours are simply types, with no linkage to taste, while the 'colours' in quantum chromodynamics, describing the interaction of quarks, have nothing to do with colour. And the quantum property of spin doesn't involve anything spinning.

All quarks have an electrical charge that is either $\pm\frac{2}{3}$ or $\pm\frac{1}{3}$. Really, we probably should say quarks have a charge of ±1 or ±2, while electrons and protons have a charge of ±3, but the charge on the electron and proton was established first.

Up, down, strange, charm, top and bottom

You can also have a point for calling the last three charmed, truth and beauty. The most familiar, and most stable, quarks are the least massive up and down, which combine in triplets to make up the atomic nucleus particles the neutron and the proton. The other quarks, while predicted by theory, are not generally observed, but can be detected when they are produced in high-energy collisions and then decay almost immediately.

In the original quark model from 1964, only up, down and strange existed (where 'strange' referred to the strangely long life of particles that were thought to include a strange quark). However, theory suggested that a fourth flavour should exist, so charm was added, despite evidence for it not emerging for nearly another ten years. The remaining pair were also conceived in theory well before being observed, this being due in particular to the fact that they require vastly more energy to produce them, which wasn't possible with the accelerators of the day.

The bottom quark was discovered in 1977. The top quark didn't turn up until 1995, not surprising as its mass, which proved not possible to predict from theory, was comparable with an atom of gold, making it an extremely heavy particle, requiring very high energies to produce it.

*Further reading: **The God Particle***

A massive question

How does the Higgs boson give mass to other particles?

Answer overleaf ➔

While you're thinking ...

The Higgs boson is named after Scottish physicist Peter Higgs,
who was one of six physicists who predicted its existence.

The Higgs boson is sometimes given the nickname 'the God
particle'. It was suggested that this was because it was so
important to physics, but in fact the Nobel Prize-winning
physicist Leon Lederman, who came up with the name, wanted
to refer to it as the 'Goddamn particle' in a book he was writing.
It was his publisher who insisted the wording be changed.

A particle was discovered in the Large Hadron Collider at CERN
in 2012 that was consistent with being a Higgs boson, and
which behaved just as such a particle should, causing a wave of
worldwide excitement about a particle that few understood.

It doesn't

In the first place, by far the largest part of the mass in atoms comes from protons and neutrons, and most of their mass reflects the energy incorporated in them to hold together the quarks that make them up. However, there is some extra mass to explain, and the explanation is somewhat messy, which is why the media struggled to explain the significance of the Higgs boson when it was discovered.

The thing that gives mass to quantum particles is not the Higgs boson, but the Higgs field. The Higgs field, like the electromagnetic field you may have come across at school, is a way of describing something that has a value at every point in time and space, a bit like a contour map dealing with the whole universe. This Higgs field was proposed to act as a kind of drag on other particles, giving them the mass they have but that the particle theory of the day said they should not have.

When scientists make the prediction of something like the Higgs field, they like to have testable evidence for it. They can't use the mass of particles, as that was why they thought up the field in the first place. But theory predicts that just as, for instance, disturbances in the electromagnetic field produce particles called photons, so disturbances in the Higgs field should produce a different kind of particle called a Higgs boson. So it was the particle as evidence for the Higgs field that so excited the physics world, rather than the discovery of the particle 'that gives other particles their mass', as it is often presented.

Further reading: **The God Particle**

Slugging it out

In traditional US physics units, what is measured in slugs?

Answer overleaf ➔

While you're thinking ...

There are some very odd units available to the scientist who doesn't want to stick with the conventional metric ones. Some are in active use because they are of a more appropriate scale for the requirement. For instance, the metric unit of area is a square metre. But particle physicists have been known to use the barn, which is 10 billion billion billion times smaller, appropriate for measuring scattering and absorption on the scale of atoms. The name comes from sayings like 'couldn't hit the side of a barn' or 'couldn't hit a barn door'.

Newspapers, meanwhile, produce their own strange standard units. For a long time in the UK, the double-decker bus and Nelson's column were favourites, while blue whales are sometimes employed as a unit of weight.

The modern scientific units are often referred to as SI units (Système International d'Unités), or the MKS system after the key units for length (metre), mass (kilogram) and time (second). Those seeking to be quirky occasionally make use of the FFF system, referring to rather more obscure units for length (furlong), mass (firkin) and time (fortnight). To achieve the three Fs, firkin is a bit of a cheat as it's actually a unit of volume, so here it refers to the mass of a firkin of water.

Mass is measured in slugs

Modern physics in all countries makes use of the metric system in which the unit of force is a newton, and the unit of mass is a kilogram. We confuse matters in common parlance by saying that something 'weighs 10 kilograms' when we really mean it has a mass of 10 kilograms. The mass of an object is a measure of the amount of stuff in it, so does not change depending on gravity. Weight, however, is the force exerted on the object by the local gravity. On the surface of the Earth, the acceleration due to gravity is around 9.81 metres per second per second, and force is mass times acceleration, so the weight of that 10-kilogram mass is actually around 98 newtons.

If we took that same object to the Moon, it would still have a mass of 10 kilograms, but its weight would be around ⅙ of its weight on Earth – around 16 newtons. To say that it 'weighs' 1.67 kilograms on the Moon makes no sense.

When it comes to the old imperial units, variants of which are used in the USA, things get more confusing. The familiar pound is a unit of weight. Something with a weight of ten pounds on Earth genuinely does only weigh 1.67 pounds on the Moon. But what about mass? This is where slugs come in. Although rarely used in the UK, the slug is the equivalent unit of mass to the pound as a unit of weight. On Earth, an object with a mass of 1 slug weighs around 32 pounds. This is because the acceleration due to gravity is around 32 feet per second per second.

*Further reading: **Physics for Gearheads***

WHAT COLOUR IS THE SUN?

QUESTION 8
Squarking photinos

What is the name of the hypothetical extension to the standard model of particle physics that brings in photinos, gluinos, selectrons and squarks?

Answer overleaf ➜

While you're thinking ...

The 'standard model' here refers to the extremely successful approach that brought together the zoo of quantum particles discovered between the 1950s and the 1970s into a single, unified approach based on six quarks, six leptons (a class of particle including electrons and various neutrinos) and four bosons, extended to a fifth with the Higgs boson.

Successful though the standard model has been, it has a number of limitations: not adequately describing neutrinos, not providing a basis for a 'theory of everything' that pulls together gravity with the other fields of nature, and not explaining dark matter or dark energy.

Physics has several standard models, notably the standard model in cosmology that requires cold dark matter and a big bang.

Supersymmetry

Sometimes given the twee contraction SUSY, supersymmetry is a way to try to pull together the very different types of particle in the standard model: bosons, such as photons of light, and fermions, such as electrons and quarks. If supersymmetry existed, it would go a long way to explaining some of the divergence between what is actually observed and the predictions of the standard model. These include unexpected particle masses, the way that the standard model doesn't allow for the main forces other than gravity to unify at high energies as it is expected they should, and the way the standard model does not provide mechanisms for combining gravity with the other forces, as do theories like string theory.

In principle, supersymmetry should be relatively easy to test because it predicts that pretty well every particle has a supersymmetric equivalent of the opposing type – so the bosons have supersymmetric partners that are fermions and vice versa. This is where the weird names come in, as the partner names are formed by changing -on ending bosons to -inos – so photons have a photino partner – and putting an s- in front of fermions, so electrons have selectron partners.

The big problem, though, is that if supersymmetry did exist, we would expect some of the supersymmetric partners to be produced with the energies available to a collider like the Large Hadron Collider; in practice no candidate particle has ever been discovered.

*Further reading: **Reality's Frame***

QUIZ 1
ROUND 6: POT LUCK

Girdling the Earth

A cable is fitted around the equator of the Earth. Assuming the Earth to be spherical, how much longer would the cable have to be if it were raised off the surface by 1 metre all the way round?

Answer overleaf ➔

While you're thinking ...

The Earth's circumference around the equator is approximately 40,075 kilometres.

In practice the Earth isn't spherical – it has bumpy bits in the form of mountains, and bulges a little around the equator because of its rotation. But it's near enough a sphere, only varying by tens of kilometres.

A 28-millimetre steel cable long enough to stretch around the equator would weigh around 130,000 tonnes.

Around 6.3 metres

You would only have to add just over 6 metres to the 40,000 kilometre-long cable to raise it a metre off the surface of the Earth all the way round. Intuitively, this feels too small a distance, but a touch of basic school mathematics will enable you to calculate any such extension in your head.

The circumference of a circle is $2\pi r$, where r is the radius, so if you expand the circle from a radius of r (we don't care what r is) to r + 1, the circumference will go from $2\pi r$ to $2\pi(r+1)$, which equates to an increase of 2π, or around 6.3. If you increase the radius by 1 metre, the circumference increases by about 6.3 metres. If you increase it by 1 kilometre, the circumference increases by around 6.3 kilometres.

More generally, if you take the cable n metres off the Earth, the radius of the circle goes up from r to r + n, so the circumference/length of the cable goes up from $2\pi r$ to $2\pi(r+n)$ – making the increase $2\pi n$, or around 6.3n.

Further reading: **Instant Brainpower**

Not that again

What scientific term with its own unit is the most commonly used noun in written English?

Answer overleaf ➔

While you're thinking ...

Although the word we're looking for is the most commonly written noun, it only comes 55th in the list of words overall. This seems odd, but there are far more nouns to choose from, so the most frequently used words tend to be articles, conjunctions, prepositions, pronouns and verbs.

This word count is based on the Oxford English Corpus, a collection of texts containing over 2 billion words on which the *Oxford English Dictionary* is based. The most common word of all, not surprisingly, is 'the', followed by the less predictable 'be', because the count is allowing all parts of 'to be', such as 'are' or 'am'. Strictly speaking this makes it a lemma count (where a lemma takes in all the forms of a particular word), not a word count.

There is at least one other scientific term with its own unit in the top 100 – it will be revealed over the page.

'Time' is the most commonly used noun in written English

Although its popularity as the number one noun is because we use 'time' in so many different ways, it doesn't prevent 'time' from being a scientific term with its own unit (the second). Similarly, 'work', which is the 87th most popular word, the 16th most popular noun *and* the 20th most popular verb, is a scientific term with its own unit (the joule, as 'work' in physics is the transfer of energy) that also has a wider general usage.

Although some physicists claim time does not exist – by which they mean, among other things, that scientific laws are often time-independent – time remains of fundamental importance in many areas of science and is fascinating because it is something we think we experience subjectively, and yet the objective measurement of time can be very different. The passage of time can vary depending on how the different bodies involved are moving, as special relativity comes into play. Similarly, the rate of time passing is dependent on gravitational fields, due to general relativity.

We might think that clock-watching and obsession with time is a modern phenomenon, but 4th-century bishop St Augustine of Hippo commented: 'What is time? Who can explain this easily and briefly? Who can comprehend this even in thought so as to articulate the answer in words? Yet what do we speak of, in our familiar everyday conversation, more than of time? We surely know what we mean when we speak of it. We also know what is meant when we hear someone else talking about it. What, then, is time? Provided that no one asks me, I know. If I want to explain it to an inquirer, I do not know.'

Further reading: **Reality's Frame**

Clock watching

What usually comes between III and V on a clock using Roman numerals?

Answer overleaf →

While you're thinking ...

The Roman numbers from one to five are rather like counting off fingers on a hand (where the thumb-to-forefinger shape of the open hand suggests a V), though the system has the added sophistication of being able to represent a number that is, for instance, one smaller than another by putting a I in front of it, so that X for ten gives us IX for nine.

The simple Roman numbers have far fewer curves than the Arabic numbers we use normally, making them particularly suitable for stone-carved inscriptions, including sundials, which may be why they were used on clock faces long after Arabic numbers became the standard.

Some early clocks had no face at all; they were restricted to chiming the hours. However, the 14th century saw the introduction of clocks with sophisticated readouts that not only told the time, but showed astronomical features, notably the lost clock of St Albans Abbey and the Wells Cathedral clock from the 1390s.

IIII is what usually comes between III and V on Roman numeral clock faces

I'm afraid you don't get a point for IV. Despite all we've been taught about Roman numerals, it is traditional for makers of clocks and watches, when providing a dial with Roman numerals, to use IIII instead of IV. No one is entirely sure why, though it does make for less confusion if you see the clock face in a mirror.

The Roman numbering system uses a small degree of positional significance. Where in our traditional Arabic numbers (which are actually Indian, but came to Europe via Arabic-speaking countries) the column a number is in signifies a factor of ten, there is no equivalent in Roman numerals. In principle they could be written in any order, with no significance at all given to the sequence, so 1642 could be IIMCDXXXX. But there is a convention that they are written in descending order, hence MDCXXXXII. Because of this convention, the Romans were then able to cut down on repetitive values like XXXX by using a system like referring to 2:45 as 'quarter to three' rather than 'three-quarters past two'. So they ended up with MDCXLII – this was only possible thanks to the ordering convention.

For whatever reason, clockmakers decided to ignore this and go with IIII, although nine on a clock face is the usual IX rather than VIIII.

Further reading: **Are Numbers Real?**

Restricted diet

What digestive system issues did the inhabitants of Edwin Abbott's two-dimensional 'Flatland' face?

Answer overleaf ➜

While you're thinking ...

Edwin Abbott had the unusual middle name
Abbott, making him Edwin Abbott Abbott.

Flatland was a novel (of sorts) written in 1884 and featuring men and
women who were, respectively, two-dimensional shapes and lines.

There were separate men's and women's doors in Flatland buildings
to prevent a woman from accidentally stabbing a man.

A conventional digestive tract needs an entrance and an exit – but to have this in a two-dimensional creature would divide the entity in two

In practice, the women of Flatland couldn't have any kind of organs at all, as they were lines and so had zero thickness. But even the men could not have a conventional digestive tract as that would need to have an entrance and an exit, which would divide a man into two totally unconnected parts.

The book isn't great as a novel, but plays around with a number of dimensional concepts, especially when an inhabitant of Flatland dreams of a one-dimensional Lineland, and is then visited by a sphere in Flatland. The sphere appears to be an extremely strange creature, starting as a point, growing to a circle and returning to a point.

There are also some aspects of social parody in Flatland, though it is no Gulliver's Travels. It is arguable that the book is more valuable for the ideas it contains, and perhaps even more so the ideas that it has inspired, rather than for any merits it may have as a novel.

Further reading: **Flatland**

Is this germane?

What was first called 'dunkle Materie'?

Answer overleaf ➜

While you're thinking ...

Although English is now the accepted international
language of science, for part of the 20th century
understanding German was an absolute essential.

The man responsible for the term
was the Swiss astronomer Fritz Zwicky.

Scientists' first ideas for names are often shaped by circumstance.
Enrico Fermi, for instance, intended to call the almost massless
particle produced in nuclear reactions the neutron, only to be
beaten to it, resorting instead to the diminutive neutrino.

Dunkle Materie = dark matter

Back in the 1930s Fritz Zwicky, then at the California Institute of Technology, discovered that a group of galaxies called the Coma Cluster were behaving strangely. There seemed not to be enough stuff in the galaxies for gravity to hold them together. Zwicky postulated there was another kind of stuff, dark matter (a direct translation of *dunkle Materie*), which also had gravitational attraction, but was invisible.

In a sense, 'dark matter' is a terrible misnomer. A bit like Fred Hoyle's term 'big bang', referring to something small that didn't bang, dark matter is not an absorber of light, as darkness suggests, but rather totally indifferent to light – not responding to electromagnetism – and so it is entirely transparent matter. For a while Zwicky's idea was ignored, but by the 1970s, there was an increasing amount of evidence that ordinary matter was simply not present in sufficient density in the universe to account for a lot of astronomical observations.

If dark matter does exist, there is about five times as much of it as ordinary matter in the universe, which makes it something we ought to know a lot more about than we actually do. A lot of work is going into researching the nature of dark matter, while a few physicists believe that it doesn't exist at all, with gravitation operating subtly differently on the scale of galaxies and clusters than it does on the scale of more familiar objects and heavenly bodies.

Further reading: **Reality's Frame**

The real alloy

Why were 40 identical lumps of platinum-iridium alloy produced in Paris in 1879 and distributed around the world?

Answer overleaf ➜

While you're thinking ...

Platinum-iridium alloys are rarely used as they are extremely expensive. However, they have the joint advantages of being very resistant to tarnishing and significantly harder than platinum alone.

Napoleon III had been removed as French Emperor just nine years earlier, and the Third Republic established.

The distribution of the 40 objects followed the signing of an international treaty in Paris in 1875, with signatories including most major European countries, including the UK, as well as the USA, Canada, India, China, Australia and about half of South America.

They were reference kilograms, defining the mass of 1 kilogram

Science needs fixed values for units to be sure that everyone means the same thing when they refer to them. Early units of measure were very vague and could vary from town to town or even individual to individual. As long as, say, a bushel of grain was about the same amount to whoever used the term, it didn't really matter whether or not it was accurate to the nearest grain. But for science, such accuracy is essential.

Initially, the main scientific units were defined by setting up standard reference versions of, for instance, a metre length or a kilogram mass. This was a good start, but any physical object, as opposed to a definition based on natural units, was in danger of varying with time. For example, a bar could change slightly in length due to temperature differences, while the reference kilograms could lose a few atoms in a scrape, or have a few added if anything rubs off on them.

Some units have proved relatively easy to link to natural measures. So, for instance, a metre is no longer defined by a piece of metal, but rather as $\frac{1}{299,792,458}$ of the distance light travels in a vacuum in a second. (We then need to define a second, which relies on the frequency of a light source.) However, it has proved harder to make a measurable unit to define a kilogram. In principle it could be determined by making it the mass of n protons (or some other particle), but the current favoured approach to replace the reference kilogram blocks is to use a device called a watt balance to make an accurate derivation of mass from the electromagnetic force required to support it under gravity.

*Further reading: **Reality's Frame***

Balloon follies

If you have a tethered helium balloon in the middle of a stopped car, what will happen to the balloon when the car accelerates forwards?

Answer overleaf ➔

While you're thinking ...

Helium is the second-lightest of the elements, and was largely formed in the big bang, even though most stars are busily converting hydrogen to helium. It is used in balloons rather than the lighter hydrogen because it is not flammable.

Helium was first discovered in the spectrum of the Sun before it was found on Earth, hence its name, from the Greek name for the Sun, Helios.

Acceleration is any change of velocity, so strictly speaking 'acceleration' includes deceleration, but, in this case, a stationary vehicle is increasing in forward velocity.

The balloon will float forwards (in the direction the car is accelerating)

A natural assumption might be that the balloon moves backwards as the car accelerates forwards away from it, but in reality the reverse happens. There are a number of ways of looking at this, but the most elegant uses Einstein's equivalence principle, met in the previous round. This says that gravity is the equivalent of acceleration in the opposite direction to the gravitational pull, and the two are indistinguishable.

This means that accelerating forwards is the same as feeling a gravitational pull backwards. This makes sense when you think about what it feels like when a plane accelerates down the runway. As the plane accelerates forwards, you are pushed backwards into your seat, as if a gravitational pull is operating towards the back of the plane.

Now think what happens to a helium balloon in air, under a downward force of gravity (the usual arrangement). The balloon drifts upwards because it is less dense than the air around it. It naturally moves in the opposite direction to the pull of gravity. So if a car accelerating forwards is like having the pull of gravity towards the back of the car, then the balloon will move in the opposite direction – towards the front of the car. And this is what it does.

Further reading: **Reality's Frame**

Trickle-down phenomenon

What is the difference between the way water rotates as it runs down a plughole north and south of the equator?

Answer overleaf ➜

While you're thinking ...

There are plenty of demonstrations on the internet showing this apparent difference, and visitors to destinations that straddle the equator often receive similar demonstrations.

The formation produced by the water as it spirals down the plughole is known as a vortex.

Vortexes (or vortices) can be extremely powerful – think, for instance, of whirlpools and tornadoes. But in the case of the plughole, the effect is very subtle.

There is no difference

The myth that is often maintained is that water will go anticlockwise down the plughole north of the equator and clockwise south of the equator. This is because of something called the Coriolis force. Like centrifugal force, the force is an effect of relative motion, rather than an actual push of one object on another.

Because the Earth is rotating, something attempting to move in a straight line will find that it appears to be pushed off course. Think what would happen if you fired a cannon sitting on a rotating roundabout. From the viewpoint of a person sitting on the roundabout, the cannonball would not fly in a straight line: it appears to be pushed off direction by a force – this is the Coriolis force.

Unfortunately, however, the Coriolis effect is very weak, and there are other forces influencing how a vortex starts, such as residual currents in the water, the exact direction in which the plug is removed, and any imperfections in the surface of the base of the basin or bath, which will almost always overwhelm the Coriolis force. This is particularly true when making comparisons of two locations in sight of each other either side of the equator, as often used in the (rigged) demonstrations. But even well away from the equator, the force is so small that it will be overwhelmed by other considerations.

*Further reading: **Inflight Science***

QUIZ 1
FIRST SPECIAL ROUND: AN ELEMENTARY MESSAGE

Uncover the coded message using the symbols of the chemical elements below:

1. Thorium Iodine Sulfur

2. Rhodium Uranium Barium Rubidium

3. Iodine Sulfur

4. Uranium Neodymium Erbium

5. Selenium Vanadium Erbium Aluminium

6. Lanthanum Yttrium Erbium Sulfur

7. Oxygen Fluorine

8. Phosphorus Lanthanum Sulfur Titanium Carbon

One point for each complete word – two bonus points for getting the whole message.

An elementary message – solution

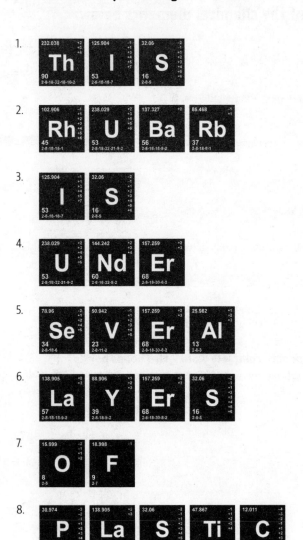

1. This
2. Rhubarb
3. Is
4. Under
5. Several
6. Layers
7. Of
8. Plastic

One point for each complete word – two bonus points for getting the whole message.

QUIZ 1
SECOND SPECIAL ROUND: TELESCOPIC KNOWLEDGE

Identify the pictured telescopes and dig deep for some telescopic knowledge:

1.

2.

3.

4.

5.

Quiz 1 | Second Special Round: Telescopic Knowledge

6. In what decade did the original 200-inch mirror telescope see first light?

7. What was the Leviathan of Parsonstown?

8. After whom is the successor visible light space telescope to the Hubble Space Telescope named?

9. In what town did William Herschel build his biggest (40-foot) telescope?

10. Where is the largest (at the time of writing) single telescope – with a 10.4-metre aperture – located?

Telescopic knowledge – solution

1. Hubble Space Telescope

2. Hale telescope (will also accept Mount Palomar)

3. Galileo's telescope

4. Very Large Telescope (will also accept Atacama Desert)

5. Newton's telescope

6. The 1940s

7. A reflecting telescope, officially the Rosse telescope, in Northern Ireland. At 72 inches it was the largest in the world for over 70 years, but rarely used because of the weather.

8. James Webb (a former NASA official)

9. Slough, England

10. The Canary Islands

QUIZ 2
ROUND 1:
ALL IN THE MIND

Picture this

What is the significance of squares, circles, stars, crosses and wavy lines in psychological experiments?

Answer overleaf ➜

While you're thinking ...

The term 'psychology' dates back to the 17th century, originally meaning the study of the soul.

The symbols were first used in this way in the 1930s.

All but one of the symbols are variants on geometric shapes; the wavy lines, which are usually in the form of three vertical lines that wiggle from side to side in parallel, look more like a symbol for water from a map.

They are the images on Zener cards, used in telepathy/parapsychology experiments

Have half a point for telepathy or parapsychology experiments and the full point for knowing that they are called Zener cards. The cards are named after Karl Zener who worked with Joseph Banks Rhine at Duke University in North Carolina in a famous series of experiments on telepathy and telekinesis in the 1930s.

Rhine asked Zener to devise the cards, which were used in experiments where one person would look at a shuffled pack of the cards and another would attempt to mentally receive a transmission of the cards' images. The idea of using the cards rather than a more easily obtained pack of playing cards was both because they were more visually distinct, so hopefully easier to transmit, and because many people have biases towards certain playing cards. (Unfortunately they also tend to have biases towards certain shapes.)

The Duke experiments had some impressive apparent successes, which were written up by Rhine in both scientific papers and popular books. Unfortunately, in hindsight, the experiments were deeply flawed, with plenty of opportunities to cheat and practically every statistical error in the book used (unwittingly) to produce apparent successes with no actual basis in reality. Their credibility wasn't helped by the fact that the original hand-made Zener cards were so thin it was possible to read the symbol from the back of the card.

The cards have become part of the cultural history of parapsychology, employed, for instance, in a humorous experiment at the start of the original *Ghostbusters* movie.

Further reading: Extra Sensory

Brain power

How much of the approximately 100-watt average energy consumption of the human body is used by the brain?

Answer overleaf ➜

While you're thinking ...

The film *The Matrix* was hugely entertaining, but its premise that human beings could be used as a source of energy was nonsense – we represent a very inefficient way of converting the chemical energy in food into a usable energy source.

A watt is a measure of power – it's a joule of energy per second. We most often come across energy for our metabolism in terms of the 'calories' on food labels. These are actual kilocalories (1,000 calories), where a calorie is an outdated measure of energy. A calorie is 4.184 joules, so a food calorie (or kilocalorie) is 4,184 joules.

When not involved in heavy exertion, a human consumes around 1,200 to 1,500 kilocalories of energy a day. That's around 5 million to 6.25 million joules. With 86,400 seconds in a day, that's around 57–72 joules per second (also known as watts).

Around 20 per cent, or 20 watts

This is a startlingly large amount of energy for a single organ, and emphasises how important the brain is to human beings that we're prepared to use up so much energy on it.

Around two-thirds of that energy consumption is used to enable the neurons to fire. The functioning of the brain is dependent on the electrochemical connections between around 100 billion of these elongated cells, which frequently have multiple connections, making as many as 1,000 trillion connections. The majority of the rest of the energy goes to what has been described as 'housekeeping' – the background biological processes, often involving transporting chemicals from place to place, that keep the brain a healthy living organ.

Lacking batteries, the energy in the body is stored in a molecule called ATP, a conveniently compact way of referring to the compound dihydroxyoxolan-2-yl methyl (hydroxyphosphonooxyphosphoryl) hydrogen phosphate. Components of our cells called mitochondria, which seem to have developed from separate bacteria in the distant past of the evolution of complex cells, create ATP from the chemical energy produced by digesting our food. The ATP molecules then act like tiny coiled springs, storing energy away in the bonds between phosphorus and oxygen atoms. The ATP is then transported to where it is going to be used, at which point these bonds are broken, releasing energy.

Experiments on rats have shown that significantly fewer ATP molecules are produced in rats' brains when they are anaesthetised, when it's assumed that the brains are primarily dealing with housekeeping, while the extra molecules produced when they are active are involved in the 'thinking' functions of the brain.

*Further reading: **The Universe inside You***

Memory module

Why is the region of the brain that is involved with long-term memory storage and access called the hippocampus?

Answer overleaf ➜

While you're thinking ...

The brain deals with a range of different memory types – for example, procedural memory, where we store our knowledge of how to do something without thinking about it (touch-typing, driving etc.), or short-term memory, which we use, for example, to remember a phone number between looking it up and dialling it.

Memory storage takes place across wide-reaching areas of the brain, but the hippocampus seems to be responsible for processing the long-term memories.

Strictly speaking we should be talking about hippocampi, as there are two of these regions, separate from each other, one on either side of the brain.

Because it is supposed to look like a seahorse

The word hippocampus is approximately a combination of the Greek words for 'horse' and 'sea monster', and the mythological hippocampus was literally a 'sea horse' with the front half of a horse joined on to a fish-like back half. It was later discovered that there was an odd-looking fish that swam upright and had a horse-like head, which naturally enough was given the seahorse/hippocampus label. (The 'hippo' part may mislead to suggest a hippopotamus, but that's because that animal's name is also partly derived from 'horse'; 'hippopotamus' combines the Greek for 'horse' and 'river'.)

When it comes to the part of the brain, the usage dates back at least to the early 1700s, with the allegation that this section looks like a seahorse. This takes a fair amount of imagination when looked at *in situ*, where the hippocampus looks more like a tadpole (it was often described in the early days as being like a silkworm). But when extracted, with its curled tail and a larger lump on top of its body, the hippocampus does indeed have something of a resemblance to the fish.

Finding the functions of the bits of the brain can be tricky, and the first guess seems to have been that the hippocampus was involved in the sense of smell – it does have a few small links to the appropriate organs – but suspicions grew when those who had their hippocampus destroyed, often in surgical errors, found it difficult to remember things that had happened. It may also play a significant role in our memories associated with location; a study has shown that London taxi drivers who store large amounts of location knowledge to pass their test appear to have larger hippocampi than the rest of the population.

*Further reading: **The Universe inside You***

A shocking experiment

In the 1961 Milgram experiment, where volunteers thought they were applying electric shocks to a subject, what percentage of participants, under extreme pressure, were reported as giving a fatal 450-volt shock?

Answer overleaf ➜

While you're thinking ...

The Milgram experiment was devised to measure how far individuals would go under pressure from authority figures. Undertaken in 1961, around the time of the Nazi war crimes trial of Adolf Eichmann, it was an attempt to assess how effective was a defence of 'I was only obeying orders'.

Participants were paid $4 an hour to take part in the experiment at Yale University, and were told they were subjects for an experiment to study memory.

In recent years there has been significant concern about the ethics of such methods, echoed in both experimental settings and some televised 'experiments', such as those undertaken by Derren Brown where participants are apparently led to perform acts they would not otherwise undertake, including a recreation of the Milgram trial.

Sixty-five per cent

Have a point for anything between 55 and 75 per cent. Although there is some dispute over the exact percentage of participants who were compliant, as attempts to reproduce Milgram's results have come up with varying figures, there seems little doubt that there is a significant effect on at least half the individuals.

The participants were told that they were acting as a 'teacher' to help another volunteer (actually an active participant in the experiment) to learn pairs of words. When the fake volunteer was tested, the teacher was to give them electric shocks on failure, to reinforce the learning. There was no actual electric shock given – the 'volunteer' was faking it – but the teacher did not know this.

Over time, the teacher was encouraged to increase the voltage applied, all the way up to a potentially deadly 450 volts. (In some experiments, the participant was told that the volunteer had a heart condition as well.) All predictions of what might happen suggested that only a tiny percentage would go all the way, despite coercion from the person running the experiment, who first cajoled and finally ordered the teacher to apply the higher and higher levels of shock. However, in reality, far more were compliant enough to go all the way and apply what they thought was a potentially fatal voltage.

More recent repetitions of the experiment have tended to be under the less controlled environment of TV shows but have continued to produce a surprisingly high level of conformance with requests. It would be interesting to discover if this reflects a shift in our idea of authority figures from academics and politicians to TV presenters and celebrities, but such a study has not been undertaken.

*Further reading: **Elephants on Acid***

The P factor

Which works better with a full bladder, short-term memory or our ability to make decisions involving self-control?

Answer overleaf ➜

While you're thinking ...

There is no definitive figure for the capacity of the human bladder, but in adults it appears to be somewhere between 400 millilitres and 1 litre.

Recent research suggests that most mammals urinate for a fairly consistent length of time – around 21 seconds. However, the standard deviation for this value was 13 seconds (meaning around 68 per cent were between 8 and 34 seconds), so in practice the figure is a fairly broad-brush one.

Human decision-making is a complex process which is rarely 100 per cent rational, being influenced by emotions and comparisons that take it far beyond the cold, calculating approach applied by economists.

Our ability to make decisions involving self-control

It isn't entirely surprising that a full bladder can result in problems with short-term memory: the need to urinate has a tendency to push other things out of the way. It also means that it is harder to pay attention, and results in an increased risk of having an accident (not just a toilet-related one).

However, apparently in contradiction, there is some evidence that despite this lack of attention and short-term memory capacity, if we have to make a decision where self-control is important, where it is essential not to make rash choices but rather to look beyond the immediate, having a full bladder appears to improve decision-making (there needs to be more research to support this suggestion, as it's based on limited experiments). The argument is that the focus given to keeping the bladder under control means that we are less likely to make a faulty quick identification of someone, or to take a financial decision that seems beneficial in the short term, but will not be good over a longer period.

Arguably, this principle (given the unfortunate name of inhibitory spillover) suggests that government ministers, who are infamous for making short-term decisions with long-term disbenefits, should undertake their decision-making processes when urgently in need of a visit to the loo.

Although the memory and decision-making findings appear contradictory, it could be argued that some people, at least, when finding it difficult to concentrate, would be less likely to make snap, risky decisions with a full bladder. As yet, though, the evidence is that self-control trumps short-term memory in this circumstance.

*Further reading: **The Universe inside You***

Enter the mad scientist

What is a ganzfeld experiment?

Answer overleaf ➜

While you're thinking ...

There is something about the name of this experiment that brings
out the mad scientist in us. If someone says to you, 'I would like
you to take part in my ganzfeld experiment', you can be sure
Igor is lurking somewhere in the background. This is probably
the legacy of Mary Shelley placing her Dr Frankenstein's initial
work in Ingolstadt in Germany (though the action also takes the
protagonist to the UK and the North Pole), reinforced by the
Germanic visuals of the 1931 James Whale movie *Frankenstein*.

The German term 'Ganzfeld' means something
like 'whole field' or 'all the field'.

The experiment involves table tennis balls.

A sophisticated experiment to detect telepathic ability

Many experiments have been attempted to detect telepathic ability, but few have been as bizarre as the ganzfeld experiments, devised in the 1960s and reaching a peak in the 1970s. The experimental format was designed to reduce sensory distraction for the subjects who act as receiver; they lay in a reclining chair with half a table tennis ball taped over each eye on to which red light was projected, while pink noise (white noise without harsh high frequencies) was played through earphones. This approach was chosen rather than total sensory deprivation, as it was thought the latter would result in the subject losing concentration.

A typical experiment involved a sender who would concentrate on a picture randomly selected from four for around half an hour, before the receiver was removed from the isolating equipment and asked which of the four pictures was used. Initial experiments showed some statistical significance, but there were a number of issues. Because of the length of time involved in the experiment, there were typically fewer than twenty trials in a study, which with a one in four chance of guessing right made statistical error highly likely.

There were also experimental errors allowing a leakage of information; in a good number of trials the receiver was presented with the actual copy of the picture that had been held by the sender. It would have been easy for the sender to accidentally leave a trace on the image that could be unconsciously picked up by the receiver. Later experiments in the 1990s with tighter controls showed no significant outcome.

Further reading: **Extra Sensory**

Oh, rats!

The 1963 Rosenthal and Fode experiment demonstrated that specially selected bright rats were better at running mazes than unintelligent ones. Why was this strange?

Answer overleaf

While you're thinking ...

The experiment involved a simple T-shaped maze and compared the rate at which the bright and unintelligent rats learned the maze in order to get to their reward more quickly.

The lab rat is usually of the species *Rattus norvegicus* (also known as the brown rat), and is closer to a domestic pet rat than a wild rat.

The rats in question were albino. Many lab rats are albino, but this doesn't appear to be for any particular reason; it's more that specific strains that have proved hardy and docile are regularly used, and some of the more popular strains, notably Wistar, Sprague Dawley and their derivatives, happen to be albino.

The 'specially selected bright rats' were not special at all, but exactly the same as the 'unintelligent rats' with which they were compared.

It's remarkable that they still get away with it, but psychologists regularly lie to their participants about what is being tested. In this case, the young scientists involved were told that the experiment was testing the ability of the two types of rat to learn, when actually the scientists were the experimental subjects, observed to see if they would indulge in cherry-picking – biasing the outcome in a particular direction by not recording everything the same way. And they did so with a will. The 'bright' rats were recorded as making successful transits of the maze up to twice as frequently as the 'dull' ones, and often made it to the prize significantly faster – at least, according to the researchers.

The experiment demonstrated that the scientists' expectations were colouring its outcome. It's not that they were consciously cheating, but because they 'knew' that one group of rats was brighter, they gave them the benefit of the doubt in a borderline case. It's also possible that they could have encouraged the bright rats more, and would have been more likely to ignore failures for a bright rat if 'something went wrong' in the experiment, such as the rat apparently being distracted from its task.

This kind of natural tendency to show bias, with no malice involved, is why it is so important where possible to undertake double-blind experiments where those undertaking the trial don't know what is being tested and so can't take a biased view. The approach is now used in all high-quality medical trials, but we are also seeing (to a degree) the approach being picked up in, for instance, particle physics.

Further reading: **Extra Sensory**

Dicing with chance

Which side of a traditional wooden die has a slightly greater chance of cropping up?

Answer overleaf →

While you're thinking ...

A traditional die (plural dice) is a cube marked on opposite faces with 1 and 6, 2 and 5, and 3 and 4. These markings are usually in the form of the appropriate number of dots, rather than numbers.

If a die is totally fair, it has a one in six (16.666 recurring per cent) chance of any side coming up.

The earliest dice were made out of bone, usually the 'knucklebone' or astragalus, and were not six-sided. For a while most were wooden cubes, but now plastic cubes dominate.

The side with 6 on

What does this have to do with psychology? Only that dice were often used in early experiments to generate a random number, and particularly in experiments to try to demonstrate telekinesis, the hypothetical ability to influence an object's movement with the mind. The most famous example was at J.B. Rhine's Duke University lab, where an experimenter threw a die a total of 52,128 times, trying to mentally influence it to land on the 6. By chance alone, with a fair die, 6 should have come up 8,688 times, but the experimenter threw 9,720. This was so many that the result was unlikely to have been down to chance alone.

To be fair to the experimenter (called Frick), he then ran the experiment again, trying to avoid getting a 6. This time 6 turned up 9,714 times. This outcome could have been predicted. The wooden die in use had the numbers marked by drilling small indentations in each side. With six indentations to the opposite side's one, the 6 side was the lightest, giving it a slightly better chance of ending up as the top face. This kind of die has a small natural bias towards 6.

Strangely, Frick's control trial was used as further evidence for the existence of telepathy thanks to some impressively twisted logic. It was argued that the attempt not to think about getting a 6 ... resulted in thinking of getting a 6. (Just try not thinking about a hippopotamus for a few seconds.) And so, it was argued, Frick's attempts not to get a 6 actually made him more likely to get a 6. In a better control experiment, a subject who didn't know about the 6 part would try to get a 1, and the die would be tested for bias with a mechanical shaker.

*Further reading: **Extra Sensory***

QUIZ 2
ROUND 2:
SCIENCE FICTION

Putting things into perspective

In the radio series *The Hitchhiker's Guide to the Galaxy* (and Douglas Adams' books) what was used in the Total Perspective Vortex device to extrapolate the whole universe?

Answer overleaf ➜

While you're thinking ...

The Total Perspective Vortex first appeared in the opening episode of the second radio series of *The Hitchhiker's Guide to the Galaxy* (titled 'Fit the Eighth'), recorded in May 1979 and broadcast on 21 January 1980.

In the episode, the totally cool Zaphod Beeblebrox is taken to the Frogstar, 'the most evil place in the Galaxy', where he is to be fed into the Vortex. The device was invented by Trin Tragula, who, fed up with his wife telling him to get a sense of proportion, built this device that extrapolated the whole of reality from a single object. The shock of seeing the whole of creation and herself in relationship to it destroyed Trin's wife's brain.

The conclusion drawn from the initial use of the Vortex was that if life is going to exist in a universe this size, the one thing it cannot afford to have is a sense of proportion.

A piece of fairy cake

According to the show's script, the inventor Trin Tragula was nagged about the time he spent staring out into space, mulling over the mechanics of safety pins, or doing spectrographic analyses of pieces of fairy cake. This is why the fairy cake was used, though the principle on which the Vortex is based is that the whole of creation could be extrapolated from anything because everything affects everything else.

While everyone else who has ever been put in the Vortex has been driven mad, Zaphod Beeblebrox, when given the chance to see himself in relation to the whole of creation, simply retorts that the Vortex just told him what he knew already – that he's a really great guy. The episode ends with him eating the piece of fairy cake.

The book version of the narrative, which features in *The Restaurant at the End of the Universe*, brings out the likely inspiration for the Vortex as a counter to the tendency of overblown TV science programmes to attempt to give a sense of wonder by showing how tiny and insignificant the Earth is compared to the Milky Way galaxy as a whole, and then to the universe. *The Restaurant at the End of the Universe* says that the Vortex gives a glimpse of the entire unimaginable infinity of creation, and somewhere is a tiny marker that says, 'You are here'.

Further reading: **The Restaurant at the End of the Universe**

Bard in space

Which Shakespeare play was the model for the classic science fiction movie *Forbidden Planet?*

Answer overleaf ➜

While you're thinking ...

Forbidden Planet, one of the more sophisticated of the science fiction films of its time, was released in 1956 starring Walter Pidgeon, Anne Francis and Leslie Nielsen. Here Nielsen plays a very straight role as the young male lead, though he would become famous for his comic acting in movies such as *Airplane!*.

The film features a large, vaguely humanoid robot, not particularly originally given the name Robby.

Unlike much of the distinctly B-movie, monster-driven science fiction films of the 1950s, *Forbidden Planet* incorporated a lot of the ideas of science fiction literature, from faster-than-light starships to alien races with a technology far beyond our own. *Star Trek* creator Gene Roddenberry notes the film as an inspiration – and it shows.

The Tempest

Although some science fiction commentators play down the connection, it's hard not to see *The Tempest* as a significant model for *Forbidden Planet*. In the Shakespeare play, the sorcerer Prospero lives on a remote island with his daughter Miranda and a deformed servant, Caliban. A group of sailors arrive on the island and the invisible spirit Ariel, commanded by Prospero, scares people off. In the film, the scientist Dr Morbius lives on a distant planet with his daughter, Altaira, and a strange servant in the form of the robot, Robby. Their idyll is disturbed by visiting spacemen and this results in attacks from a mysterious disembodied entity, a 'monster of the id', summoned by Morbius' unconscious mind using the ancient high-tech machinery on the planet.

One of the most striking aspects of the movie is the use of so-called 'electronic tonalities' that replace the usual background music. Even today, the assorted beeps, screechings and hummings are striking – all the more so because they were created before the development of synthesisers, using electronic circuits put together by the avant-garde composers Bebe and Louis Barron. It has been said that the producers made use of this source to avoid paying guild fees, though it's not clear if this is true.

With a relatively sophisticated plot and some impressive special effects, notably the vast underground Krell city on Altair IV, *Forbidden Planet* was probably the most influential science fiction film until *2001: A Space Odyssey*, which came out twelve years later.

Further reading: **Ten Billion Tomorrows**

A frightening future

What futuristic weapons of war did H.G. Wells describe in his 1914 book *The World Set Free*?

Answer overleaf ➔

While you're thinking ...

Wells had something of a reputation for spotting military concepts early. He described tanks in his 1903 story *The Land Ironclads*, though he dismissed any original thinking, claiming he was inspired by Leonardo da Vinci's wooden tank-like devices. Wells also predicted fixed-wing warfare in his 1908 novel *The War in the Air*, where bombs were dropped from planes just five years after the Wright brothers' maiden flight.

Wells produced some extremely readable science fiction, most notably his short stories and the novels *The War of the Worlds*, *The Invisible Man* and *The First Men in the Moon*. (Although *The Time Machine* proved a huge inspiration to others, it's less effective as a science fiction narrative.) However, in the 20th century, Wells turned out a number of books, including *The World Set Free* and *The Shape of Things to Come*, which, while containing very interesting ideas, tended to be turgid political futurology, rather than readable novels.

After a fairly bumpy start in academia, Wells spent a couple of years at the Normal School of Science (now part of Imperial College, London), studying biology. He would later receive a BSc in zoology as an external degree from the University of London.

Atomic bombs

In *The World Set Free*, Wells describes a war that pits Britain, the United States and France against Germany and Austria. In this future world of 1956, radioactivity is used to produce electricity and as the war progresses, new and terrible weapons making use of nuclear energy are deployed – Wells calls these 'atomic bombs'.

Science fiction isn't really about predicting the future, it's more about seeing how humans react to the challenges thrown up by science and technology – real or imaginary. But in books like *The World Set Free*, Wells was intentionally speculating about the future. The accuracy of his vision was quite remarkable when you consider it would be another 20 years before the chain reaction that made nuclear weapons possible was discovered, and that as late as 1933 Ernest Rutherford, who made a huge contribution to atomic and nuclear science (and whose moustache was the equal of Wells'), would say, 'The energy produced by the breaking down of the atom is a very poor kind of thing. Anyone who expects a source of power from the transformation of these atoms is talking moonshine.'

Another of Wells' books that comes closer to futurology than science fiction is *The Shape of Things to Come* from 1933, which was made into the ponderous film *Things to Come* in 1936. Although by then the threat from Germany was increasingly obvious, it must have seemed prophetic that the book predicted a Second World War with Nazi Germany breaking out in January 1940.

Further reading: **Ten Billion Tomorrows**

Game on

Which invention from *Star Trek* inspired John Carmack to produce first-person shooters *Wolfenstein 3D*, *Doom* and *Quake?*

Answer overleaf ➜

While you're thinking ...

The *Star Trek* franchise has included six distinct TV series, with a seventh at the planning stage at the time of writing, and three generations of films. The original series began broadcasting in 1966.

John Carmack is a US programmer who was joint founder in 1991 of the id Software company. His speciality has been sophisticated graphics engines underlying the games and at the time of writing Carmack is Chief Technology Officer of Oculus VR.

Carmack was arrested aged fourteen when he broke into a school in a failed attempt to steal Apple II computers.

The holodeck

Although the technology used in *Star Trek* played a central role, the series rarely featured totally original ideas. The transporter, for instance, responsible for the idea of beaming something up (though the phrase 'Beam me up, Scotty' was never used) is simply a variant on the matter transmitter. This had appeared in science fiction as early as the 1870s, and became standard fare in the pulp science fiction of the 1920s and 1930s, though many versions required both a transmitter and a receiver, where *Star Trek*'s transporter was designed to avoid the effects expense (and screen time) of taking a shuttle down to the surface of a planet.

The holodeck, which first appeared properly in *Star Trek: The Next Generation* in 1987 (it featured to a degree in the 1970s animated series), is less common in other science fiction, although there have been stories, such as Ray Bradbury's 'The Veldt' from his collection *The Illustrated Man*, where what appears to be a sensory experience becomes all too real. The holodeck was a virtual reality environment which combined holographic projection with tractor beams and force fields to make it possible to interact physically with the virtual world. (The *Star Trek* writers never satisfactorily explained how the holodeck appeared to be so much bigger than it actually was.)

When John Carmack saw the holodeck, he was captivated. At the time computer games could not offer anything like the first-person experience of being in an environment in which the player could freely move around. With increasing subtlety from *Wolfenstein 3D* through to *Quake*, Cormack developed software engines that would attempt to duplicate the holodeck experience visually. This approach has become standard for many of the biggest computer games in history.

Further reading: **Masters of Doom**

Dino duel

Who would win in a fight between a *T. rex* and Godzilla (the 2014 movie version) ?

Answer overleaf ➔

While you're thinking ...

Known as Gojira in the original Japanese, but anglicised as Godzilla, the giant creature was supposed to be a prehistoric monster, revived from beneath the ocean by nuclear radiation. The name is a combination of 'gorilla' and *kujira*, the Japanese for 'whale'.

The first Godzilla film came out in 1954 in Japan, reaching the USA in 1956, with sections of US actor Raymond Burr edited in to provide some exposition. There have been around twenty Godzilla films in total.

The *T. rex* became a movie star almost three decades earlier, appearing in the 1925 film *The Lost World*, animated by Willis O'Brien.

The *T. rex*

The obvious winner should be Godzilla, which over the span of the movies has got bigger and bigger. In the original 1954 film, Godzilla was around 50 metres tall. By the latest (at the time of writing) 2014 film, the monstrous dinosaur had reached a towering 106 metres. That's about two-thirds the height of Blackpool Tower, a third the height of the Shard and just over half the height of the Washington Monument. By comparison, a T. rex was a puny twelve metres long and more like five metres tall. The largest dinosaur discovered to date, found in Argentina, was around 20 metres tall, but that was like a larger version of an Apatosaurus (the dinosaur formerly known as the Brontosaurus).

With Godzilla between ten and twenty times bigger, it seems strange that the T. rex would win a fight – but this is because the first step Godzilla took, his legs would snap like twigs. The bigger an animal is, the wider its legs have to be in proportion to its body. Compare, for instance, the relative thickness to the size of the body of the legs on a mouse and an elephant. This is because the weight of animal goes up with its volume. So if we double all the dimensions, its weight goes up by a factor of $2 \times 2 \times 2 = 8$. But the strength of the legs goes up with their cross section, so doubling the dimensions would increase the leg strength by $2 \times 2 = 4$. With the ten or twenty-fold increases, Godzilla would not stand a chance.

Further reading: **Ten Billon Tomorrows**

QUESTION 6
Ancient spacemen

Of what was Lucian of Samosata's 2nd-century AD book *True History*, featuring a voyage to the Moon, a parody?

Answer overleaf ➔

While you're thinking ...

Lucian of Samosata was born around AD 120 to 130 and in the confusing mix that the Roman Empire could offer was a Roman, living in Syria, who spoke and wrote in Greek. (The ancient Samosata was located near the current Turkish city of Samsat.)

Although there is much dispute over when science fiction began, it's possible to consider Lucian's story, unusual for the period for being a long narrative written in prose, a predecessor of science fiction. The classification is not easy as the space travellers get to the Moon using the unlikely vehicle of a whirlwind, making the story more like fantasy.

There was at least one other long narrative prose work featuring a voyage to the Moon before Lucian's – *The Wonders Beyond Thule* by Antonius Diogenes, but this work is now lost.

The Odyssey

A good modern equivalent of Lucian's book seems to be *Bored with the Rings*, the Harvard Lampoon take on *Lord of the Rings*. It seems that Lucian intended consciously to parody *The Odyssey* and similar works that portrayed clearly fictional events as if they were fact. We tend to simply think of Homer's epic poem about the voyages of Odysseus, written as early as the eighth century BC, as a dramatic piece of fiction, but Lucian is known to have commented that he was surprised that the likes of Homer supposed they could get away without people noticing they were lying. From this, it seems that the ludicrously unlikely happenings in *True History* (not to mention its title) are a clear attempt to point out this situation.

Although, as mentioned overleaf, *True History* does not have any reasonable scientific mechanism for getting to the Moon, neither do many of its successors. And there is no doubt that the story features a whole range of ideas that would become common in science fiction, from alien life and interplanetary warfare to telescopes and air that has turned liquid. The fact that it is a parody does not, of itself, stop it from being science fiction; this is a mix that the likes of Douglas Adams and Robert Rankin have since made popular.

Further reading: **Ten Billion Tomorrows**

Teleporting troubles

What happened to Vincent Price's screen brother in a 1958 B-movie featuring teleportation?

Answer overleaf ➔

While you're thinking ...

Price, born in 1911, was a master of the B-movie horror film. To a whole generation he was probably best known as the creepy voice on the Michael Jackson song 'Thriller', but his horror work, notably in his frequent collaborations with director Roger Corman, was unrivalled.

At the start of the film, in which Price plays François Delambre, his brother André (they are French Canadian) is found horribly mutilated, his head and arm crushed in a hydraulic press.

The word 'teleportation' is first used in the 1931 book *Lo!* by researcher of the strange Charles Fort. In using the term, Fort envisaged moving objects by the power of the mind, but by the time the Vincent Price movie came out, the term was applied to an altogether more scientific kind of matter transmission.

He got mixed up with a fly

Science fiction often gives us scientists whose lives are put at risk or even destroyed by their search for the truth – never more so than André Delambre, who invents a matter transmitter (unlike a *Star Trek* transporter, this is a classic matter transmitter, swapping the contents of two enclosed chambers). This was a popular theme at a time when radio had transformed communication from place to place, and the ideas of quantum physics seemed to suggest that the distinction between waves like radio and the particles that made up matter was an artificial one.

In the movie *The Fly* (remade in the 1980s by director David Cronenberg), a fly gets into the matter transmitter with the scientist, their atoms are mixed up and we end up with a man with a fly's head and leg (in place of one of his arms), and a fly with a human head and arm. Realistically, this was a highly unlikely scenario. After all, the person being transmitted was in a chamber full of air – how could the device be able to avoid mixing human and air, but not human and fly? We also have no logical explanation for why the fly's head was blown up to human size and the human head reduced to fly size.

However, we shouldn't be too harsh. This was, after all, primarily a horror story using science fiction trappings, and at the time was considered impressively shocking.

*Further reading: **Ten Billion Tomorrows***

Tremendous tech

What do the science fiction devices the ansible and the Dirac transmitter have in common?

Answer overleaf ➜

While you're thinking ...

Science fiction contains a whole range of fictional devices from time machines to faster-than-light drives. Some will probably never become reality. Others tend to be outpaced by real technology – the communicators in the original series of *Star Trek*, for example, seem feeble compared with a smartphone.

The ansible was devised by Ursula K. Le Guin and the Dirac transmitter, named after the real-life British physicist Paul Dirac, by James Blish.

It is quite common for different science fiction writers to come up with their own solutions to problems presented in a science fiction universe. For instance, when it came to voyaging to the stars, some used vast 'generation ships' where hundreds of people would live their whole lives on journeys lasting hundreds of years, while others devised a range of mechanisms for getting around the light-speed barrier.

They are both instantaneous communication devices

Blish's Dirac transmitter first turns up in 1954 in a short story called 'Beep', then plays a major role in the oddly named novel *The Quincunx of Time*. The problem with any instant transmitter is that it has to somehow get around the speed of light being the maximum speed for information to be transmitted. When Blish wrote the stories, positrons – antimatter electrons – were hot science news, so he picked up on them, just as Isaac Asimov would for the 'positronic brains' in his robots, more because they were trendy than because there was a scientific reason for choosing them.

Having said that, Blish did pick up the oddity of this new antimatter, predicted by Dirac's equation, to wonder if in some way positrons and electrons could be linked, so that one could influence the other, however far apart they were. There was less of an explanation of how an ansible worked, but it had interesting limitations meaning it could only send short text messages, and had to be based on a planet with a significant mass.

The ansible, described in Le Guin's novel *The Dispossessed*, was merely a long-range communicator. But Blish realised his Dirac transmitter, using relativity, could send a message back in time. Blish has an overblown version of this, where every transmission ever made is compressed into a bleep heard at the start of each message. This wouldn't happen – but it does reflect the strange time-travelling capability of the instant message.

*Further reading: **Ten Billion Tomorrows***

QUIZ 2
ROUND 3: MISCELLANY

QUESTION 1
Metal miasma

When can you breathe a metal (and survive)?

Answer overleaf ➔

While you're thinking ...

Any metal *can* be breathed if heated to gaseous form, but
this is not recommended if you want to stay alive.

The word 'metal' comes from the Latin *metallum*, which referred
to both what we would call a metal and a mine or quarry.

In English, the metals were originally only gold, silver, copper, iron,
lead, tin and their alloys, but the term came to include a wider class
of elements that had similar physical properties to the original six.

When you are an astronomer

Astronomers are arguably the most obtuse of scientists, who insist on using their own terminology, even when it doesn't fit with the rest of the scientific community. So, for instance, where the official scientific unit of distance is a metre, astronomers rarely use this, instead opting for either light years – the distance light travels in a year – or parsecs, based on parallax, the variation of angle a distant object produces when seen from the two extremes of the Earth's orbit. (They tend to choose between these units depending on the nature of the measurement, breaking a fundamental rule of universal units that they should not depend on the method of measurement.)

However, this misuse pales into insignificance when put alongside the astronomers' definition of a metal, which is anything other than the main products of the big bang, hydrogen and helium. The argument for this division could be that metals are things that are made in stars. So, for instance, the oxygen we breathe was the product of various fusion reactions in ancient stars. However, this runs entirely contrary to the way all other scientists think of metals. It's also magnificently inconsistent. For example, although helium was, indeed, a product of the big bang, a lot is also made in stars – the Sun, for instance, is busily converting hydrogen into helium.

To make matters even worse, there was some lithium (which is a metal) made in the big bang. We don't have a clear first usage, but the astronomers' misuse goes back at least to a thesis by US astronomer Nancy Roman, written shortly after the Second World War, referring to 'metallic lines' in the spectra of stars.

*Further reading: **The Cambridge Illustrated History of Astronomy***

QUESTION 2
Roman rounding

How many fewer letters does it take to write 2000 than 1999 in Roman numerals?

Answer overleaf ➔

While you're thinking ...

Early number systems were based on tally marks, which could be notches on a piece of wood or bone, one notch per object counted. The interesting thing about tally marks is that they can be used without counting. You can pair off tally marks and objects to see if the right number are present without knowing how many objects there are.

Roman numerals include tally marks, such as I, II, III for one, two and three, and some letters, which represent the names of numbers.

While C and M for 100 and 1,000 are name-based (*centum*, *milia*), V and X for five and ten seem to be advanced tally marks (V based on the thumb-to-forefinger shape of the open hand; X as a pair of joined Vs). D (500) is debatable. It could be a half-thousand (*dimidium mille*) or it could be a cross between a V and a circle, sometimes used to mark the hundredth occurrence in a tally.

Five (MM versus MCMXCIX)

You can also score for eight or fourteen. As is described elsewhere (page 98), the Roman numbering system has a small degree of positional significance. Numbers are usually listed in descending order, but this order can be varied a little by putting, say, a smaller number in front of a larger one to indicate 'reduce by' – so, for instance, four can either be IIII or IV.

In this case, the most compact form for 1999, and the one we would probably use today (for example on TV programme copyright notices) is MCMXCIX, where the 'cut down by' approach is used to reduce 1,000 to 900 (CM), 100 to 90 (XC) and ten to nine (IX). However, the Romans themselves seem not to have been hugely enthusiastic for writing nine as IX – for example, doorway number 29 in the Coliseum at Rome is labelled XXVIIII rather than XXIX. This would give us MCMXCVIIII, giving the reduction of eight.

It was also certainly perfectly acceptable in Roman times not to use contractions at all, hence coming up with the magnificent MDCCCCLXXXXVIIII and that massive reduction of fourteen. It may be unwieldy, but it requires the least interpretation. For some reason, other possible combinations of contraction and none, such as MCMLXXXXVIIII or MDCCCCXCIX, seem not to have been used.

Further reading: **A History of Mathematics**

Scientific headcount

To the nearest 100, how many individuals worldwide, both in universities and out, would have described themselves as scientists in 1820?

Answer overleaf ➜

While you're thinking ...

While there is some dispute over when the role of scientist in the modern sense began, there can be no doubt that Isaac Newton, born on 25 December 1642 (which would be 4 January 1643 on our calendar) was a working scientist.

Newton was the second Lucasian Professor of Mathematics at Cambridge, a position that would be held by as many scientists as mathematicians, with the likes of George Airy (an astronomer), George Stokes, Paul Dirac and Stephen Hawking (all three physicists) following Newton in the chair.

The Royal Society, founded in 1660, and the Royal Institution, founded in 1799, have been host to huge numbers of scientists with, for instance, fifteen of the Royal Institution's scientists having won Nobel Prizes.

Zero

Were there scientists in 1820? Without doubt. Exactly who was the first scientist is open to huge debate; it's possible to go for anyone from, say, Archimedes, through the 9th-century Arabic scholars, via medieval figures like the 13th-century friar Roger Bacon to Galileo, who pretty well everyone would agree was a scientist. But not one of these people would have called himself a scientist. They were all natural philosophers.

They would have been aware that they were involved in science – which essentially referred to knowledge based on observation of the world – and of topics like physics and mathematics (maths, confusingly, would have included astronomy because unlike the physics of the time, astronomy involved numbers). However, there was no word 'scientist'.

It was felt by the 19th century that 'natural philosopher' was both clumsy and imprecise, as philosophy had come to have a tighter meaning than its original far-reaching 'love (or study) of learning' – and because philosophers thought science was beneath them. Other options such as 'man of science', 'sciencist', 'sciencer', 'scientician' and 'scientman' had been dabbled with, but in 1834, rather controversially, at a British Association for the Advancement of Science meeting, 'scientist' was proposed. Those present had considered 'savant' (meaning 'a learned person'), but this was thought to be 'assuming', so instead devised 'scientist' by analogy with 'artist', pointing out that 'economist' and 'atheist' were already acceptable terms. The meeting found that this was 'not generally palatable' – but despite this initial resistance, 'scientist' stuck.

Further reading: **Science: A History**

Crazy counting

Where can 9 + 5 = 2?

Answer overleaf ➔

While you're thinking ...

We tend to assume that numbers are base ten, so 9 + 5 = 14, as this is the way that we usually count, but it's easy enough to use, for example, base two – binary, as used in computers – or, for that matter, any other base. We have only settled on ten because of the ten digits on our hands.

Arguably, using base five, counting on one hand rather than two, would have been more effective, as we could then count up to 30 on our fingers and thumbs. We would count up on the left hand in the normal way, but for a full hand would then close the left hand and extend the thumb of the right. With the second full left hand we extend the right-hand forefinger – and so on, all the way to 30. In effect, the hands become a simple abacus.

The Sumerians and the Babylonians counted to base 60, which is very flexible, as 60 can be divided by 1, 2, 3, 4, 5, 6, 10, 12, 15, 20 and 30. It might seem unmanageable, but bear in mind we still use this approach for minutes and seconds, and in some measurements of angles.

On a clock face

It may seem a bit of a cheat, but there is an entirely respectable branch of mathematics dealing with 'clock arithmetic', more formally known as modular arithmetic. In this type of arithmetic, on hitting a certain value – the modulus – the numbers reset and start again. We know, for instance, that five hours after 9am is 2pm – so 9 + 5 = 2. You can imagine clock arithmetic to be the result of rotating a hand around a clock. The approach can be done with any modulus, not just twelve, and is useful in representing any cyclic action, not just the action of a clock.

There are all kinds of applications of modular arithmetic. At its simplest, it's often used as a check in long code numbers like some bank account numbers, where numbers are issued with a key number that acts as a check value using modular arithmetic.

More importantly, many encryption algorithms, including the RSA algorithm often used to secure computers for safe transactions, make use of modular arithmetic in checking a key, in a rather messy process known as modular exponentiation, which makes use of the remainder (the final value on the clock) of dividing one number raised to the power of a second number by a third number.

*Further reading: **Are Numbers Real?***

QUESTION 5
Devilish science

Which law of physics was Maxwell's demon devised to break?

Answer overleaf ➜

While you're thinking ...

Physicist Richard Feynman defined the laws of physics as follows:
'There is [...] a rhythm and a pattern between the phenomena of
nature which is not apparent to the eye, but only to the eye of
analysis; and it is these patterns which we call Physical Laws.'

The Maxwell in question was Scottish physicist James Clerk Maxwell,
who among other things took the first colour photograph and produced
the equations describing the behaviour of electromagnetism.

Maxwell's demon is an example of the thought experiment,
or *Gedankenexperiment*, beloved of theoretical physicists
like Albert Einstein to test a theory by setting up a
hypothetical but impractical experiment in the mind.

The second law of thermodynamics

The existence of a challenge to this law is dramatic. Physicist Arthur Eddington once remarked, 'If someone points out to you that your pet theory of the universe is in disagreement with Maxwell's equations – then so much the worse for Maxwell's equations. If it is found to be contradicted by observation – well these experimentalists do bungle things sometimes. But if your theory is found to be against the second law of thermodynamics I can give you no hope; there is nothing for it but to collapse in deepest humiliation.'

The second law of thermodynamics is, roughly: 'heat moves from a warmer to a cooler body in contact' or 'in a closed system, the level of disorder [entropy] stays the same or rises'. The demon comes in by imagining a simple experiment with two boxes of identical gas, each containing a mix of hot (fast-moving) and cool (slow-moving) atoms. There is a door between the boxes. If we just leave the door open, over time things should stay roughly the same as atoms randomly move between the boxes. But then we add the demon, an imaginary creature that can open and shut the door as atoms head towards it. The demon does this in such a way that only fast atoms can go left to right, and only slow atoms right to left. So after a while, one box is hot, the other cold. As one gets hotter and the other cooler, heat is moving from hot to cold. And the more all the hot atoms are in one place and the cool ones in another, the more order increases.

There are a number of counters to the demon that require energy to be put into the system (so the law doesn't apply), but it has never been entirely resolved.

*Further reading: **Reality's Frame***

Jurassic jaunt

What are the chances of being able to build a park featuring living dinosaurs within the next 20 years?

Answer overleaf ➜

While you're thinking ...

The original book *Jurassic Park* was written by Michael Crichton.

Since the first *Jurassic Park* film, our ideas about the appearance of dinosaurs have changed considerably – for example the Velociraptors should have feathers – but the movie series has kept the original appearance, in part probably because making the predators more realistic would render them less scary.

Many of the dinosaurs portrayed in *Jurassic Park* are not from the Jurassic era, but from the Cretaceous.

Absolutely certain (or 100 per cent)

In fact, the chances are that you have seen (or will see) a living dinosaur today, without even venturing to a park. Because birds are dinosaurs. For a long time most of us have realised that birds and dinosaurs are related, but they aren't just a distant relative. Birds are the direct descendants of the ancient dinosaurs, and if you look at the scientific classification of birds, you will find that they come within the clade (group of organisms) Dinosauria, just like old friends such as *T. rex* and *Diplodocus*. Birds *are* dinosaurs.

By contrast, various animals that we tend to lump in with dinosaurs aren't from the same clade. So, for example, none of the flying reptiles called pterosaurs, or the marine ichthyosaurs or plesiosaurs are dinosaurs. (To be really irritating, most of the pterosaurs weren't pterodactyls – there's only one species of these.)

To underline this relationship, palaeontologists now tend to refer to what most of us just call dinosaurs as non-avian dinosaurs, to avoid bringing in the living members of the group.

Further reading: **The Tyrannosaur Chronicles**

Strange substance

Is glass a solid?

Answer overleaf ➔

While you're thinking ...

The familiar three states of matter (solid, liquid and gas) have been joined by at least two others: plasma and the Bose–Einstein condensate.

Pitch, used in tarmacadam roads, despite appearing to be solid, is a liquid. The longest-running lab experiment of all time involves a lump of bitumen pitch that has been monitored at the University of Queensland since 1927. In that time nine droplets have fallen, the most recent in 2014. You can watch a live feed at www.thetenthwatch.com. It's not awfully exciting.

Medieval glass panes tend to be thicker at the bottom than the top. This has often been explained as the result of the glass having flowed downwards extremely slowly under the pull of gravity.

Yes, glass is a solid

Sometimes it's the obvious answer that is correct, and that's the case with glass – it really is a solid. For some time it was misrepresented as being a very slow-flowing liquid, like tar but even slower, and the medieval glass window panes were given as an example of the way that the glass very gradually flows under the pull of gravity. In reality, however, medieval glass was already uneven – they did not have modern mechanisms like the float process to achieve a smooth, flat sheet of glass. When a glazier put a piece of glass in place, the thickest edge was usually put at the bottom for stability – the glass didn't flow into that state. We have Roman glass that is far older than the medieval panes, but this shows no sign of having 'flowed under gravity'.

One of the reasons that glass doesn't seem like many other solids is that it doesn't have a crystalline structure – it is amorphous, a random mix of molecules, rather than the more familiar regular structures. But it has the essentials of a solid of not flowing to fill a container, holding its own external shape.

Another confusing material is custard. A thick mix of custard powder and water is clearly a liquid, but put it under pressure and it solidifies – you can pick it up, or even walk on it. This property of changing state under pressure is known as being thixotropic.

*Further reading: **The Universe Inside You***

QUESTION 8
Draw data

In a lottery picking 6 from 49 balls, which is most likely to come up: 1, 9, 15, 31, 38, 44; 1, 2, 3, 4, 5, 6; or 1, 2, 4, 8, 16, 32?

Answer overleaf ➔

While you're thinking ...

The UK's main national lottery game Lotto originally required a match of 6 balls from 49, making the chances of winning 1 in 13,983,815. In 2015, the ball count was raised to 59, reducing the chances of winning to 1 in 45,057,474. There have been so many rollovers as a result that the number may be modified again.

At the time of writing, of the balls between 1 and 49 in the UK Lotto, 20 has been drawn the least with 217 draws and 23 the most with 286 draws. (The numbers above 49 have only been drawn between four and eight times so far.) This tells us nothing about what will be drawn next.

At the time of writing, the longest a ball (number 50) had gone without being drawn in the bi-weekly draw was over three months.

None of them – they are all equally likely

Our perception of randomness rarely matches reality. For example, if asked to write down a series of numbers between 1 and 9, picked randomly, we tend to put in far fewer repeats than actually occur. Similarly, if we see a pattern in a set of numbers, it seems less random than a set of apparently unconnected numbers. There is an assumption that the pattern has a cause, making it less likely to be a random occurrence.

In practice, any six balls are just as likely to be drawn as any other six. So 1, 2, 3, 4, 5, 6 is just as likely to turn up as 1, 9, 15, 31, 38, 44. If 1, 2, 3, 4, 5, 6 were drawn, there would no doubt be an outcry, and it's easy to see why. There is only one way to make 1, 2, 3, 4, 5, 6 – but there are lots of ways to make a sequence without any obvious pattern – here's another: 3, 7, 8, 19, 20, 44. So it's much more unlikely that 1, 2, 3, 4, 5, 6 will come up than 'any apparently pattern-free sequence'. However, this doesn't make any difference between the chances of 1, 2, 3, 4, 5, 6 and 1, 9, 15, 31, 38, 44 occurring.

More interesting, in a way, is 1, 2, 4, 8, 16, 32 – which also has a pattern, doubling on each value. What it reminds us of is that there are plenty of ways the numbers could be drawn where there is some apparent pattern. So though 1, 2, 3, 4, 5, 6 is extremely unlikely (like any single sequence), the chance of a draw coming up with a suspicious pattern is far higher than we might expect.

Further reading: **Dice World**

WHAT COLOUR IS THE SUN?

QUIZ 2
ROUND 4: SPACE

Solar blues

What colour is the Sun?

Answer overleaf →

While you're thinking ...

Stars come in a range of colours from deep reds to brilliant
blues. (This excludes oddities such as black holes and
brown dwarfs, which are not-quite stars that fall between
a large gas planet like Jupiter and an active star.)

The most common stars in our galaxy are red
in colour, specifically red dwarfs.

Star colours reflect their surface temperatures and are classified
on a confusing scale that runs O, B, A, F, G, K and M from the
hot O to the relatively cool M. The Sun is a class G star.

White

When we ask someone to draw the Sun, whether a child or an adult, they tend to draw a yellow circle, which is odd because generally speaking the Sun either appears too bright to look at or a reddish colour as it heads for the horizon. However, there is a kind of logic as the white light of the Sun has some of the blue stripped out of it, scattered by air molecules and producing the blue colouration of the sky. The result is that the Sun itself appears to be more yellow than it actually is, hence the astronomers' confusing labelling it a yellow dwarf (it is neither yellow, nor particularly small). In reality, the Sun is white, producing the familiar white light that can be split into a spectrum by a prism.

A subtler error is that made by the TV show *QI* in suggesting that the Sun is blue or turquoise in colour. This reflects the way that the Sun puts out a whole range of energies of photon (or wavelengths if you prefer to think of your light in a wavy fashion), but doesn't produce equal intensities of all the different possibilities in the visible spectrum. The light peaks in the blue and then green ranges, hence the suggestion that sunlight is blue or turquoise. Unfortunately, 'colour' is not a simple reflection of the distribution of photon energies because it is a subjective human response, not an objective scientific measure of energy distribution. Our eyes, just like the Sun, don't respond equally to different energies (or wavelengths if you prefer). What we see in the brain's response to the combination of inputs from the various sensors in the eye is a white colour for direct sunlight.

*Further reading: **Bright Earth***

QUESTION 2
Greeks in space

What is the astronomical significance of the Greek gods Hermes, Aphrodite, Ares, Zeus and Cronos?

Answer overleaf ➜

While you're thinking ...

Aphrodite's role in this context was originally taken by
a combination of Hesperus and Phosphorus.

The Ancient Greeks had a considerable collection of gods, of
whom four out of five in this list were major gods of Olympus.
The odd one out was Cronos, father of Zeus, who was a Titan, a
member of the order of gods that predated the Olympians.

The word astronomy comes from the combination of
the Ancient Greek words for 'star' (*astron*) and 'to name'
(*nemein*), reaching English via the Latin *astronomia*.

They were the names of the five true planets that featured in the Ancient Greek solar system

So, Hermes was the name for Mercury, Aphrodite was Venus, Ares was Mars, Zeus was Jupiter and Cronos was Saturn. We now use the Roman names, their equivalent gods having been substituted. As well as these five, the ancients considered the Sun (Helios in Greek) and the Moon (Selene) to be planets in the sense of being 'wandering stars' that were thought to orbit the Earth.

The reason that Aphrodite's role was originally taken by a combination of Hesperus and Phosphorus was that Venus, closer to the Sun than us, is usually only visible close to sunset or sunrise. This led to an assumption that there were two separate planets, Hesperus (which became Vesper in Latin), seen in the evening, and Phosphorus (Lucifer in Latin), visible in the morning. The Greeks eventually realised that they were the same body (as the earlier Babylonian civilisation already had), renaming the planet Aphrodite.

When extra planets were added to the solar system, invisible to the naked eye and so not observed in ancient times, the Latin version of the naming convention was extended. Uranus was the Latin version of the early Greek sky god Ouranos, who predated the Titans, and in some mythology was the father of the first generation of Titans. This was followed by Neptune, the Roman version of the Greek sea god Poseidon (back to the Olympian generation) and temporarily Pluto after the god of the underworld (Plouton in Greek).

The odd one out is the Earth, with names originating in ancient words for 'ground'. The planet Earth was Terra in Latin and Gaia in Greek.

*Further reading: **The Cambridge Illustrated History of Astronomy***

WHAT COLOUR IS THE SUN?

Live long and prosper

Where was the planet Vulcan once thought to orbit?

Answer overleaf ➔

While you're thinking ...

Mr Spock's home world in the *Star Trek* universe was called Vulcan.

The French astronomer Urbain Le Verrier was the first to suggest
that the planet Vulcan existed, in the middle of the 19th century.

Le Verrier had form when it came to the discovery of planets:
he was co-discoverer of the planet Neptune with British
astronomer John Couch Adams – though exactly how much
of the glory belongs to each is disputed to this day.

Inside the orbit of Mercury

There was indeed a fictional planet Vulcan that was the home of Mr Spock in the original series of *Star Trek*, which it was decided should orbit the (real) star 40 Eridani A. This triple star system is around sixteen light years from Earth and was an early suggestion for a possible star to have a habitable planet, but as yet no planets have been discovered to be orbiting it.

The 19th-century Vulcan was less exotic. Rather than an inhabited planet orbiting a distant star, Le Verrier thought that there was a planet that orbited the Sun closer than Mercury, sufficiently small and near to the Sun to escape being discovered by astronomers. It was named after the Roman god of fire, from whose name we get the word 'volcano'.

The hypothetical planet Vulcan was not created as a science fiction home for aliens – it would have been far too close to the Sun to be inhabited – but rather to explain the orbit of Mercury, which did not behave quite as Newton's gravitational equations predicted. This oddity was explained in the 20th century by Einstein's general theory of relativity, which deviates from the predictions of Newton's mathematics in conditions near massive bodies like the Sun.

This is different from the idea of a CounterEarth, orbiting the opposite side of the Sun to Earth, which first cropped up in early Ancient Greek astronomy as Antichthon, and has since regularly featured in fiction, but for which there has never been any evidence.

*Further reading: **The Cambridge Illustrated History of Astronomy***

A distant light

Light has been crossing the universe for 13.8 billion years. So what is the radius of the observable universe?

Answer overleaf ➜

While you're thinking ...

Light travels at around 186,000 miles per second, or, to be precise, 299,792,458 metres per second.

For convenience, astronomers often give distances in light years. In one year, light travels one light year.

In practice, light didn't start crossing the universe until the universe was about 300,000 years old, but for the purposes of this exercise we can assume that we can see light that has been travelling for the vast majority of the lifetime of the universe.

Around 46.5 billion light years

It's not a trick question, and yet despite the universe only being 13.8 billion years old, giving light just under 13.8 billion years to head towards us from the furthest point we can see, the distance of the most distant objects is far greater than 13.8 billion light years away.

The explanation is expansion. During its lifetime, the universe has expanded hugely. And not only is it expanding, but the further away you get, the faster that expansion is. So the most distant things we could see are around 46.5 billion light years away. We would see such distant objects as they were around 13.8 billion years ago – it's just that during that time, they have moved further and further away to reach their present distance.

We have no idea how big the total universe is – just that the segment that we can observe is this kind of size. Nor are we saying that we are currently able to see to this distance – the observable universe is an ultimate limit at any particular point in time, but this doesn't mean that we have the technology to observe objects at that distance. At the time of writing, the most distant observed object is the galaxy GN-z11, from which the light has been travelling towards us for 11.4 billion years. Many articles misinterpret this to say that GN-z11 is '11.4 billion light years away', but its proper distance is much further. It is difficult to establish exactly how far, as calculating it depends on a range of assumptions about the nature of the universe, but the galaxy is likely to be at least 32 billion light years away.

*Further reading: **The Cambridge Illustrated History of Astronomy***

WHAT COLOUR IS THE SUN?

Not macho

In astronomy/cosmology, what is a WIMP?

Answer overleaf ➔

While you're thinking ...

The term 'wimp' for a weak and feeble person had a first recorded use in 1920, but this seems to have been a one-off. Exactly where it came from is uncertain, though it has been suggested it could be derived from 'whimpering'. The word became common usage in the 1960s.

No one has ever seen or directly detected a WIMP, yet they may be very important in astronomy and cosmology.

Rather confusingly, the term WIMP was first used in science in the mid-1980s at exactly the same time as it began to be used in computing to stand for 'windows, icons, mouse and pointer'. Because such graphical user interfaces are now ubiquitous, the term has pretty much disappeared from computing, but is still common among cosmologists.

A candidate for dark matter – stands for Weakly Interacting Massive Particle

Give yourself one point each for dark matter and for 'Weakly Interacting Massive Particle'. As we saw on page 102, dark matter was first predicted back in the 1930s and became more widely accepted in the 1970s as a material that exists in the universe in large quantities – there being perhaps five times as much of it as ordinary matter – but which only interacts with ordinary matter gravitationally and so cannot be detected by telescopes. This matter causes, for instance, galaxies to survive when rotating too fast for the gravitational attraction of just the ordinary matter in them to hold them together.

Over the years there have been a range of theories on the nature of dark matter, including well-known but hard-to-find particles like neutrinos and hypothetical, never-detected particles called axions. But the best-supported theory suggests that dark matter is made up of WIMPs. As yet there is no evidence for the existence of WIMPs other than the dark-matter gravitational effect.

Some physicists argue that the WIMP model is ridiculously oversimplified. After all, the standard model of particle physics describing ordinary matter contains seventeen different particles. We are assuming, based on no evidence whatsoever, that WIMPs are a single particle type, even though they are far more abundant than ordinary matter. It is entirely possible that dark matter, like ordinary matter, has a considerably more complex collection of fundamental particles and may even have totally invisible and undetectable equivalents of light travelling between objects made of dark matter particles.

*Further reading: **Reality's Frame***

It's in the stars

What is Sagittarius A*?

Answer overleaf ➜

While you're thinking ...

Sagittarius is one of the twelve constellations used to mark the segments of the sky known as the zodiac. Sagittarius is an archer, usually represented as a bow-wielding centaur. In practice, the constellation looks like nothing more than a meaningless web of stars.

Sagittarius contains none of the sky's most notable stars, so has none of the well-known named stars. Its brightest star is Epsilon Sagittarii.

The name 'Sagittarius A*' refers to one of several features in the Sagittarius constellation, collectively known as Sagittarius A.

A powerful radio source thought to be a supermassive black hole at the centre of the Milky Way

Black holes themselves were an early concept that emerged from Einstein's general theory of relativity, and though they cannot be observed directly, their presence has increasingly been inferred from their influence on visible matter around them, often producing radiation as matter accelerates towards the black hole.

It is increasingly thought that galaxies regularly have particularly large black holes at their centre, and it seems likely that such black holes can play a major role in the way that galaxies form. While an 'ordinary' black hole forms from a star with just a few times the mass of the Sun, these central black holes are 'supermassive', with mass that is thousands or more times that of the Sun, having accumulated far more than the matter from a single star.

The radio source Sagittarius A* is well positioned near the centre of our Milky Way galaxy, and from the behaviour of visible objects relatively close, it appears to have a mass of around 4.3 million times that of the Sun. There is nothing present in visible light, but this is likely to be due to the large amount of dust and gas that is in the way. Sagittarius A* was first spotted in 1974, but it's only relatively recently that it has been identified as the Milky Way's central black hole.

*Further reading: **Gravity's Engines***

Ground control to Sputnik 1

In what year was the first artificial satellite, Sputnik 1, launched?

Answer overleaf ➜

While you're thinking ...

We're used to big names behind the US space programme, such as Wernher von Braun. But few have heard of Sergei Pavlovich Korolev, the man behind the USSR's R7 launcher and Sputnik.

Sputnik 1 was 58.5 centimetres across and weighed 83 kilograms, 61 per cent of which was its batteries.

The satellite remained in orbit for around three months, although the transmitter gave out after 22 days.

Sputnik 1 was launched in 1957

What is, perhaps, most remarkable is that this date manages to seem both very early and surprisingly late. At around 60 years before the time of writing, and just twelve years after the end of the Second World War, it was a long time ago. Yet the idea that the first satellite was launched only twelve years before the first man set foot on the Moon is a remarkably short timescale – especially when there has been so little advance in manned space flight since 1969.

It was less than a month after the launch of Sputnik 1 that the USSR achieved another first, turning a stray dog into a star. On 3 November 1957, Laika was boosted, alive, into orbit and into the public eye in a half-ton satellite. Sadly the cute-looking, terrier-like mongrel died within a few hours as Sputnik 2 overheated, but it showed that dire warnings that life could never survive in space were incorrect. This spurred the USA on, but things did not start well as the first attempts to rival Sputnik failed on take-off. To make matters worse, in 1961, the USSR managed another breakthrough with Yuri Gagarin's Vostok 1 mission. This made him the first human to make a true space flight, for a single orbit, on 12 April 1961.

This series of events is arguably why the USA then succeeded in getting a man on the Moon by the end of the 1960s, spurred on by a John F. Kennedy speech. The achievement was nothing to do with scientific research and everything to do with political (and military) pride.

Further reading: **Final Frontier**

QUESTION 8
Twinkle, twinkle

What is the distance to the nearest star?

Answer overleaf ➔

While you're thinking ...

There are somewhere in the region of 100 billion to
400 billion stars in our galaxy, the Milky Way, alone.

Distances in space are often measured in light years (the distance light
travels in a year), or parsecs (around 3.26 light years, a measure based
on parallax), but you are welcome to answer in kilometres if you prefer.

Stars output light as a result of nuclear fusion. Their
visible intensity is a combination of the distance
away and the absolute brightness of the star.

One hundred and fifty million kilometres, 0.00001522 light years or 0.00000467 parsecs – the distance to the Sun

Have a point for give or take 10 million kilometres. Nobody is surprised that the Sun is a star, but we often forget that it is the nearest star, going instead for Proxima Centauri at around four light years (you don't get a point for that, sadly).

Measuring the distance of stars is a non-trivial exercise and involves rather more approximation than astronomers usually admit. Relatively close stars can be measured geometrically, using parallax. When we hold up a finger and view it through the left and then right eye, the finger seems to move because the eyes are located in a different position. The closer the finger is, the more it appears to move. This same approach is used to measure the distance to stars, but instead of using the separation of two eyes, the measurement is taken on either side of the Earth's orbit, with a 300 million-kilometre separation.

For the vast majority of stars, however, the parallax technique doesn't work. Astronomers rely on 'standard candles' – particular types of body that typically have very similar absolute brightness, so by comparing two examples, it is possible to deduce the distance from their relative brightness. Typical standard candles for closer measurements are variable stars, where classes of star tend to be of very similar brightness, while for greater distances, extremely bright objects like quasars can be used.

*Further reading: **Reality's Frame***

QUIZ 2
ROUND 5:
QUANTUM STUFF

Colour me confused

Whereabouts in the spectrum does the colour magenta come?

Answer overleaf ➜

While you're thinking ...

Isaac Newton identified the seven colours of the rainbow: red, orange, yellow, green, blue, indigo, violet – but these are arbitrary divisions. The spectrum contains vast numbers of colours.

The magenta colour code in the 256 × 256 × 256 colour palettes often used on computers is #FF00FF.

The primary colours many of us are taught at school – red, yellow and blue – aren't primary colours at all. The true primary colours are those of light – red, green and blue.

Nowhere – magenta isn't on the spectrum

Not every visible colour is part of the rainbow spectrum of visible light; instead, colours like magenta are the result of the way that our eyes deal with combinations of colours. The Red, Green and Blue (RGB) colour code #FF00FF indicates that magenta is effectively white light with green missing – a mix of full intensity red and blue.

Those 'primary' colours we learned in school – red, yellow and blue – are actually the secondary colours magenta, yellow and cyan. Each corresponds to the absence of one of the primaries from white light, as pigments work by absorbing certain colours. So printers work with 'CYMK' for 'cyan, yellow, magenta and black'.

At the fundamental level, sight is a quantum interaction. Each photon of light has a specific energy, which is the equivalent of its wavelength or frequency – it defines the photon's 'colour'. A white light source emits a whole mix of different energies, and when we see something as a particular colour, it absorbs certain energies of photon because the electrons in the atoms of the object are able to make a quantum leap of a suitable size, absorbing the energy of a photon. When we see an object as magenta, it is absorbing the green photons, but re-emitting the other energies, so we receive photons from the rest of the spectrum without the green segment.

Magenta is a simple 'missing' colour; there are also compound missing colours where a wider set of energies of photon are absorbed. For example, there is also no brown in the spectrum, which involves more of the green and blue being absorbed than the red, but some absorption in all three primary colours occurs.

*Further reading: **The Universe inside You***

WHAT COLOUR IS THE SUN?

Subatomic shortcut

How does quantum tunnelling make a smartphone less forgetful?

Answer overleaf ➜

While you're thinking ...

This technology is not limited to smartphones, but turns up in most information and communications technology (ICT) devices.

Quantum tunnelling, which involves a quantum particle getting through a barrier that should be too high or strong for it to pass over, appears to take no time to get from A to B. This is not an issue for special relativity, as it's a result of the quantum particle not having a precise location.

Quantum tunnelling is one of the reasons we're alive. Even in the immense temperatures and pressures in the Sun, the positive charged particles repel each other too strongly to get close enough to enable the fusion reaction to take place. It's only because tunnelling enables them to overcome this barrier that the Sun can generate the energy that keeps us alive.

Tunnelling is used in a phone's storage (flash memory)

There are a number of possible answers here, but by far the most significant is flash memory. A phone's storage doesn't lose data when the battery runs out. (The same goes for a modern laptop with a solid state drive.) Such 'flash memory' uses one of the most bizarre aspects of quantum physics to tuck data safely away.

Flash memory was devised by Toshiba in the early 1980s. It was expensive and slow to read, so was only used in the basic input-output system (BIOS) of the computer, where key information to enable the operating system to boot was stored. However, by the 1990s, a new generation of flash memory technology was being deployed, initially as the memory cards for digital cameras and more recently, as the storage became smaller, cheaper and faster, as a robust replacement for hard disk drives.

Each bit of information stored in the flash memory is held on a special transistor called a floating gate. In this, the floating gate is an insulated section which influences whether or not current can flow through the transistor – it acts as a switch, which is how the 0 or 1 of the memory is stored. Because the floating gate is insulated from its surroundings it doesn't lose its charge when the power is switched off. But there is a problem: how to change the value, to set or unset the switch?

The solution is quantum tunnelling, enabling an electron to cross the barrier without moving through the insulated space in-between. Without this quantum tunnelling the memory would not function.

*Further reading: **The Quantum Age***

WHAT COLOUR IS THE SUN?

Blinded by the light

Why is the speed of light in a vacuum denoted as 'c'?

Answer overleaf ➜

While you're thinking ...

The speed of light in a vacuum is not the only speed of light
– light travels more slowly when passing through a medium,
as the photons interact with the particles in the matter – but
'c' is the ultimate speed of light, the universal speed limit.

It's often said that nothing can travel faster than light, but strictly
speaking the limitation is that nothing can accelerate through light
speed – hypothetical particles called tachyons could exist that always
travel faster than light, but are unable to drop down below light speed.

Another get-around on the light speed limit is quantum tunnelling,
during which the particle appears to have no transit time. This means
that if a particle tunnels through a barrier, then covers a distance at
high speed, its average over the total distance can be greater than
light speed. Such 'superluminal' experiments are limited to very
short distances and are impossible to make use of practically.

Because it's either 'constant' or 'celeritas'

Have a point for either, as the exact derivation seems uncertain. Initially, the speed of light (or more precisely its velocity, which encompasses speed and direction) was typically represented like any other speed. We wouldn't expect a special symbol for the velocity of sound, or of a car on a motorway. It is typically represented as 'v' with an optional subscript if there is more than one velocity, such as v_1 or v_{light}.

This was still the case when Maxwell developed his equations for electromagnetism that identified light as a special case of an electromagnetic wave, which needed to travel at a special, specific speed to exist – so Maxwell and Einstein each referred to the speed of light simply as 'v'.

However, two factors seem to have come together to make a move to a different symbol necessary. One was that the equations of special relativity often require a comparison of the speed of light and the speed of a moving body. A common factor in these equations, itself given the name gamma (γ), takes the form

$$1 - \frac{v^2}{c^2},$$

which would have been fiddlier than $1 - \dfrac{v_b^2}{v_1^2}$.

More importantly, however, the nature of the speed of light in a vacuum as a universal constant with significantly greater importance than just a passing velocity became clear. Hence the first reference to the use of 'c' suggests that it came from 'constant'. However, there were plenty of other universal constants that had already been identified, so perhaps a more satisfactory explanation is the Latin word *celeritas*, meaning speed. After all, it's hard to argue against as a characteristic of light.

*Further reading: **Reality's Frame***

Strange sea

What quantum particle was originally conceived as a hole in an infinitely deep sea of negative energy electrons?

Answer overleaf

While you're thinking ...

A hole in this sense is simply a missing particle. A similar concept is used in solid state electronics, where the flow of electrons in one direction is sometimes more usefully considered as a flow of 'holes' in the opposite direction.

The infinite electron sea was dreamed up by the British physicist Paul Dirac as part of his attempt to produce an equation that described the behaviour of an electron that took into account both quantum mechanics and special relativity.

Dirac was famously limited in his verbal communications. At a lecture he was giving in the USA, when it came to time for questions, a member of the audience said, 'I don't understand the equation in the top right-hand corner of the blackboard.' Dirac did not say anything, resulting in an uncomfortable pause until he was prompted, when he replied, 'That was not a question, it was a comment.'

The positron

You can also have a point for 'anti-electron'. Dirac spent a considerable time attempting to devise an equivalent of Schrödinger's quantum-mechanical equation that would adequately describe the behaviour of an electron travelling at relativistic speeds. Eventually he did so – but there was a price attached. It required electrons to be able to have positive or negative energy. This seemed to suggest that electrons would be able to give off an infinite amount of energy, plunging down to lower and lower negative energy states.

As a fairly desperate fix to this problem, Dirac suggested that every possible negative energy state was already full – that the universe contained an infinite sea of negative energy electrons, filling up all the negative energy states, so that the electrons we observe are forced to occupy states with positive energy.

However, it's always possible for an electron to receive energy from an incoming photon and to jump up to a higher energy level. If one of these negative energy electrons did so, it would leave behind a gap. And Dirac was able to show that a missing, negatively charged, negative energy electron would appear exactly the same as a present, positively charged, positive energy particle. His model predicted the existence of a positive equivalent of an electron, something that would not be found for a number of years. Ironically, when it was first found it was announced at a lecture in Cambridge, where Dirac was based – but he was on sabbatical in the USA and didn't hear about it. More than a year would pass before Dirac was made aware of the proof of the existence of the positron, the first antimatter particle.

Further reading: **Reality's Frame**

Light mill

A Crookes radiometer is like a light bulb containing paddles on a rotating spindle. The paddles are painted white one side and black the other. Exposed to light, the paddles rotate under the influence of the light (there is no mechanism). Do they go black side first or white side first?

Answer overleaf ➜

While you're thinking ...

The device was sometimes known as a 'light mill', drawing a parallel with the action of a watermill or windmill, where an impact on paddles or sails causes a spindle to rotate.

William Crookes was an English scientist (with an impressive moustache) working in the latter half of the 19th century. He specialised in vacuum tubes – glass tubes with a lot of the air sucked out, which were like early versions of the cathode ray tube used in old TV sets.

A radiometer was originally a device for measuring angles (from *radius*), but Crookes used it to describe his invention in the *Proceedings of the Royal Society* in 1875, and after that it became used more generally for a device to measure rays, or radiation.

White side first

The interesting thing about the Crookes radiometer is that it does exactly the opposite of what it should do if it were to demonstrate what it is often described as showing. Quantum theory tells us that light should exert pressure – because although photons don't have any mass, they do have momentum, the 'oomph' that makes things move. This is most dramatically demonstrated in space with solar sails, which are large-scale sails used to pick up light pressure from the Sun to move a space vessel. Although solar sails have yet to be deployed practically, they have been demonstrated to work.

If light pressure were the mechanism of the radiometer, when it is placed in a bright light, the paddles ought to rotate black side first. This is because the black side tends to absorb light, while the white side reflects it. Because momentum is one of the physical properties that is conserved, we would expect that there will be more of a push on the white sides, where the light is bouncing back off, than on the black sides.

In practice, though, the paddles rotate the other way. There is a light pressure effect, but there is a much bigger push in the other direction. This is because the black sides warm up as they absorb light. Although the bulb has some of its air removed, to make it easy for the paddles to be rotated, it doesn't contain a vacuum. Air molecules near the black sides will be warmed up, moving more energetically. So they will bounce more off the black sides, pushing them away: this starts the rotation.

Further reading: **Universe inside You website**

Suspicious beards

What quantum device was nearly scuppered when a scientist couldn't get security clearance because his referees had beards?

Answer overleaf ➔

While you're thinking ...

Because the scientist in question failed to get security clearance to work on his own project when it moved from commerce to defence, he was told he was no longer allowed to read his own notebooks.

Although beards retained their Victorian popularity longer in the UK, in part because of the bearded 20th-century kings Edward VII and George V, in the USA they rapidly fell out of fashion as the 'clean cut' look became the only acceptable one. By the 1950s, outside academia a beard was considered a sign of rebellion and dangerousness.

Despite the widespread acceptance and fashionable nature of beards in the 21st century, the US establishment has continued to avoid them – they remain highly unusual in US politics.

The laser

Although the Nobel Prize for the laser went to Charles Townes, Nikolay Basov and Alexandr Prokhorov, the main developers were Americans Gordon Gould and Theodore Maiman. Gould developed the basics of the gas laser and came up with the name, while Maiman built the first working laser, which used a ruby as its lasing material.

Gould could well have beaten Maiman to the first device if he had not had so much trouble with security clearance. To get help with his idea, Gould had taken it to the company Technical Research Group, which specialised in military contracts. Gould helped TRG win a contract with the US Department of Defense's Advanced Research Projects Agency. In fact, so impressive was Gould's pitch that when TRG asked for $300,000 to develop the laser, they were awarded $1 million.

Unfortunately, with the military contract came the requirement for workers to have security clearance. Gould had dabbled with socialism in his youth and had lived with his wife before they got married, both of which were considered to make him a security risk. Another of the factors that got in the way of his clearance was that two of his referees had beards, which was considered to be a sign of being a subversive.

For some time before he got his clearance, Gould was not allowed to read his own notes, nor was he allowed to enter the laboratory where his team was working. It is entirely possible he would have built a working laser first, were it not for this restriction. In practice, Theodore Maiman, at Hughes Corporation, achieved the first working laser on 16 May 1960.

Further reading: Ten Billion Tomorrows

Beam me up

What spooky quantum phenomenon is necessary for quantum teleportation?

Answer overleaf ➔

While you're thinking ...

Matter transmitters might be the stuff of science fiction, but quantum teleportation is a real effect that is practised in laboratories around the world.

In quantum teleportation, one or more properties from one quantum particle are transferred to another quantum particle, which can be located remotely. These properties define the nature of the particle, so, in effect, the second particle becomes a duplicate of the original.

Quantum teleportation is an essential process for building effective quantum computers. These are computers that use quantum particles like electrons or photons as the 'bits' in the computer. Because each particle can be in a superposition of states – simultaneously existing as probabilities of having different values – it can, in effect, carry out multiple calculations simultaneously, making a working quantum computer a much sought-after goal.

Quantum entanglement

It was Albert Einstein who first called quantum entanglement 'spooky'. It does appear very strange – probably the most surprising in all of what is, after all, a very surprising area of physics. When a pair of particles become entangled (there are a number of mechanisms achieving this), they become linked in a way that means they are no longer individual entities. This means that one particle can have a direct influence on the other, however far apart the particles are, as long as they remain entangled.

Take, for instance, a property called spin. When measured this will always come out as 'up' or 'down' – but before measurements the particles are in what's called a superposition of states – effectively each particle is both up and down. If we make a measurement of one particle it has to be either up or down – and the moment that measurement is made, the state of the other particle becomes the opposite. This happens instantaneously, whatever the distance.

Einstein suggested that the particles must already 'know' what the outcome of measurement will be, even though we have no way of discovering that value before measurement. However, many experiments have been undertaken since Einstein's time and all support the view that entanglement is real and instant communication occurs. Despite many efforts to do so, there is no way to use this effect to send information faster than light, as the values are totally random. There is no way of controlling, for instance, whether the first particle is spin up or spin down – hence no way to use entanglement to send a message between the two locations.

*Further reading: **The God Effect***

Scientific cobblers

What was it about physics that made Einstein say that, if it were true, he would rather be a cobbler?

Answer overleaf ➜

While you're thinking ...

Albert Einstein was, without doubt, one of the greatest scientists in history. Born in 1879 and dying in 1955, he saw the whole of physics turned on its head by relativity and quantum theory, to both of which he would make a major contribution.

Einstein was one of the last great physicists who could make major breakthroughs on his own and without the support of an academic institution. In 1905 he wrote four remarkable papers, one of which won him the Nobel Prize, despite being employed as a clerk in the Swiss patent office at the time.

The 'cobbler' remark came in a series of letters exchanged between Einstein and one of his close friends, the physicist Max Born. These letters, a mix of social and scientific observation, provide a fascinating commentary on the development of modern physics.

The basis of quantum physics on probability and randomness, rather than solid, fixed values

According to the new quantum physics, for which Einstein had very much helped lay the foundations, quantum particles did not have clear properties like position or energy, but rather had a range of probabilities for these values that would only become an actual value when they were measured. So, for instance, a single photon did not have a clear position. And while it was possible to predict the probability of a radioactive particle decaying, there was no way to predict an actual time when it would occur.

As early as 1909, when quantum theory was still in its infancy, Einstein remarked that he was 'discomforted' by the role of randomness in quantum behaviour. This had become a firmer dislike by 1924, when he wrote to Max Born, 'I find the idea quite intolerable that an electron exposed to radiation should choose of its own free will, not only its moment to jump off, but also its direction. In that case, I would rather be a cobbler, or even an employee in a gaming house, than a physicist.'

A couple of years later he wrote another letter to Born that crystallised his view to provide one of his most famous quotes (though often given in a variation). Einstein remarked: 'Quantum mechanics is certainly imposing. But an inner voice tells me that it is not yet the real thing. The theory says a lot, but does not really bring us any closer to the secret of the "old one". I, at any rate, am convinced that He is not playing at dice.' God, he suggested, does not play dice. However, the universe appears to disagree.

*Further reading: **The God Effect***

QUIZ 2
ROUND 6: CHEMISTRY

In your element

What is the element with the highest atomic number found in nature?

Answer overleaf ➔

While you're thinking ...

The atomic number is used to distinguish elements from each other and corresponds to the number of protons in the nucleus of the atom. The lightest element, hydrogen, has a single proton and the atomic number 1. There are variants of elements called isotopes with varying numbers of additional neutron particles in the nucleus, but as long as the number of protons remains the same, an atom remains of the same element.

Elements with very large numbers of protons tend to be unstable – there is simply too much repulsive positive charge crammed into the nucleus for the atom to stay in one piece. Over the years, a range of these high atomic number elements have been created artificially, such as 103, lawrencium, and 114, flerovium, but these elements do not exist in nature.

The most basic elements, hydrogen and helium, were created in the big bang, as were small amounts of lithium. But the rest of the elements – and plenty more helium and lithium – had three possible sources. The elements up to iron (atomic number 26) were created in the fusion reactions in stars, while many of the heavier elements were produced when old stars exploded in supernovas. A few elements, such as technetium and promethium, are the result of fission reactions from the breakdown of unstable radioactive elements.

Plutonium – atomic number 94

The answer that is often given is uranium (atomic number 92) – and certainly we were not aware of plutonium until it was created in an accelerator. This was in 1940, when it was produced by Glenn Seaborg and his colleagues at the University of California, Berkeley. Seaborg was an enthusiastic element hunter, for which he won the Nobel Prize in Chemistry, and has a number of elements named after him and his laboratory locations: americium, californium, berkelium and seaborgium – which became the first element named after a person who was still alive.

Later in the Second World War, far larger quantities of plutonium were produced as part of the Manhattan Project and used in both the Trinity test explosion at Alamogordo, New Mexico, in July 1945 and the bomb dropped on Nagasaki in August 1945. Plutonium itself is a silver-grey metal that continually gives off a small amount of heat from its radioactive decay.

However, despite uranium often being given the laurels as the naturally occurring element with the highest atomic number, plutonium has now been detected in small quantities in nature, such as in the remnants of natural nuclear reactors in Africa. It also now comes in a semi-natural form, where naturally occurring uranium has been converted into plutonium by neutrons emitted by nuclear fission reactions – this is now the most common source of plutonium in nature.

Further reading: **Periodic Tales**

Superhero substance

How might denatonium benzoate save a child's life?

Answer overleaf →

While you're thinking ...

In the early science fiction books and comics it was common to invent new elements with strange powers (but nowhere to go on the periodic table) – perhaps most famously, the 'element' (though it later became an alloy) kryptonite that in its various forms had all kinds of effects on Superman. Denatonium might sound like a made-up element, but despite having an element-like name it is, in fact, an organic compound.

This colourless, odourless substance was first produced by accident at a Scottish pharmaceutical firm that was attempting to produce a new anaesthetic for dentists.

Denatonium benzoate, usually appearing under a brand name, is credited with saving a large number of people, particularly children, from suffering and death.

Often marketed as Bitrex, denatonium benzoate is extremely bitter and so is added to products that might otherwise be accidentally consumed

Denatonium (which is a lot easier to say than phenylmethyl-[2-[(2,6-dimethylphenyl)amino]- 2-oxoethyl]-diethylammonium, its full name) benzoate is the most bitter substance known. The usual measure for bitterness is quinine, the active substance in tonic water, yet the tongue needs 1,000 times more quinine to taste the bitterness than it does denatonium benzoate.

We identify bitter as one of the five key tastes, and it is a flavour that we are particularly sensitive to, perhaps because so many poisonous substances produce a bitter response, although with practice we can come to appreciate a touch of bitterness on the palate. Familiar bitter favourites like coffee and beer (and tonic water) all tend to cause aversion in children, but to be more popular as we age and can appreciate taste combinations where the natural reaction is 'avoid'.

Sometimes denatonium benzoate is used in very obvious applications, for example added to the poisonous but sweet-tasting antifreeze compound ethylene glycol that otherwise might seem appealing for children to drink. It's also often added to rodent poison, as, unlike us, rats don't seem to have the same aversion to bitterness. At other times, the compound is applied to turn a product into a special variant for commercial reasons. So, for instance, 'denatured alcohol' is often produced by adding denatonium benzoate to alcohol, which makes it useless for drinking, but it can still be used in fuels and for other purposes. Finally, denatonium benzoate has a role in habit modification when it is painted on to nails to prevent them from being bitten.

*Further reading: **Royal Society of Chemistry – Compounds***

The mysterious ore

The Curies worked through tonnes of pitchblende to get tiny amounts of radium and polonium, but what compound is the main constituent of pitchblende?

Answer overleaf

While you're thinking ...

There are large amounts of the ore pitchblende in Africa, the Americas and Europe. Marie and Pierre Curie were working on ore taken from the Joachimsthal region on the German–Czech border.

As a result of the discovery of radium, which became popular in everything from toothpaste to glow-in-the-dark clock dials, large quantities of pitchblende were mined and it became a popular source for coloured glazes for (slightly radioactive) pottery and tiles.

The dangers of radium were first exposed when workers who painted luminous dials began to suffer sores and cancers around the mouth. The female workers used to lick their brushes to make a sharp point, leaving radioactive residue around their mouth. Over 100 would die from radium poisoning.

Uranium dioxide

The modern name of pitchblende, uraninite, is rather more of a give-away. Uranium dioxide was used in pottery glazes, notably a bright orange-red range from the US company Homer Laughlin called Fiestaware through to the early 1960s, with a break during the Second World War when all the uranium dioxide was requisitioned for the war effort.

This simple compound remains the single biggest source of radioactive materials for the nuclear industry. When uranium is enriched, usually necessary to a degree for reactors, and much more for weapons, the oxide is converted to the volatile uranium hexafluoride, which makes it easier to separate the more radioactive isotopes in gas centrifuges. But it is then converted back to uranium dioxide for storage and use in reactors.

Pottery and tiles making use of uranium oxide glaze are still slightly radioactive, but are generally considered not to be a risk unless someone is in long-term contact with them, or if the glaze is damaged so the compound can leak out into food and drink.

Perhaps the most impressive household use of uranium oxide predates the Curies' work. In Victorian times it was sometimes used to colour glass, giving a yellow-green shade that glows bright green under ultraviolet light. We know that this 'vaseline glass', so called because of a colouration that is reminiscent of petroleum jelly, was already being made in the 1830s as the then Princess (later Queen) Victoria was presented with vaseline glass candlesticks in 1836.

Further reading: **Royal Society of Chemistry – Compounds**

Don't get stung

Why is the traditional treatment for relieving the pain of bee and wasp stings different?

Answer overleaf ➜

While you're thinking ...

Although different in many ways, bees and wasps are both
of the suborder Apocrita within the order Hymenoptera.
Ants form a third member of the suborder.

The most familiar wasps are deceptively similar in lifestyle compared
to bees, living in a nest with a single queen laying the eggs and
many workers supporting her. However, there are many more
species of wasp where the insects are solitary with no collective
nest – many of these wasps lay their eggs on or in other insects.

There is a huge number of different bee species – so far
around 20,000 have been discovered. Although more live
socially than do wasps, not all bee species form colonies.

Because one injects an acid venom and the other an alkali substance

Probably the best-known difference between bee and wasp stings is that most bees' stings are barbed and tend to be left behind in the victim, killing the bee, while wasp stings are smooth, meaning they can be used repeatedly – seeming to lead to more aggressive behaviour. A side effect of its multi-sting capacity is that a wasp uses less of its venom than a bee, which can deliver between four and twenty times the quantity of a typical wasp (very large wasps like hornets have a larger payload). But the treatment is more to do with the venom's chemical properties.

The good news, as far as the traditional treatments go, is that bee venom is acidic, containing formic acid (also used by ants), while wasp stings are alkali. This leads to the traditional treatment of bee stings with a neutralising alkali, such as sodium bicarbonate, while wasps tend to get treated with vinegar.

In reality, any relief from either of these is more psycho-logical than practical. Apart from anything else, the sting of a bee or wasp produces a very small orifice in the human skin, with the venom injected beneath the surface. Pouring on a sodium bicarbonate solution or vinegar is unlikely to have much impact on the injected substance. More importantly, the venoms are not just an acid or alkali. In the wasp, for instance, the pain comes from a mix of amines such as sero-tonin and histamine, plus peptides known as wasp kinins.

Even if it were possible to get direct access to the venom, and the pH were the sole cause of pain, there is only a tiny quantity of venom present, so the treatment would push the pH in the opposite direction, producing its own negative effects.

*Further reading: **Venoms of the Hymenoptera***

Atomic shades

What colour is oxygen?

Answer overleaf ➜

While you're thinking ...

The air we breathe is around 78 per cent nitrogen and 21 per cent oxygen. The remainder is, perhaps surprisingly, mostly the noble gas argon (about 0.93 per cent), with the next biggest component carbon dioxide at around 0.04 per cent.

Oxygen is an element with atomic number 8 – so it has eight protons in the nucleus. By far the most common isotope is oxygen-16, which has eight neutrons, but oxygen also has stable isotopes with nine and ten neutrons and a host of short-lived radioactive isotopes.

Although oxygen was rare as a separate element on Earth when the planet first formed, it is a common element in the universe, coming third after hydrogen and helium, forming around 1 per cent of the content of the universe by mass.

Oxygen is light blue

As oxygen gas appears pretty well colourless, this seems unlikely, but it's probably the best answer to the question. Liquid oxygen has a clear pale-blue to violet colour; there is no question about its colour in this form, and arguably it's only when we see it in a liquid form that we can sensibly assess the element's colour. We tend to think of it as a gas because that is its natural state on Earth's surface – it liquefies at around –183°C. But there is nothing inherently natural about the gaseous state. The solid form has a similar colour to the liquid. (There is a red form of the solid, but this structural oddity only occurs under high pressures.)

Even as a gas, the usual description of 'colourless' is open to debate. Both nitrogen and oxygen in the air undergo a process called Rayleigh scattering. Incoming white light hits the gas molecules, is absorbed and is re-emitted in different directions. The higher the energy of the photons, the more likely they are to produce scattering – so blue light is scattered more than reds and yellows, making the sky blue.

Traditionally this scattering is described as an effect: the gas in the air isn't blue, it just causes a blue effect as a result of scattering. But the distinction is arbitrary. The colour we perceive of a substance is dependent on how it absorbs and re-emits various colours of light. While the exact mechanism is different for scattering and pigmentation, in both cases the colour we see is not an inherent property of the object, but the result of an interaction of light with the object at the atomic level. So, arguably, even oxygen as a gas could be said to be blue.

Further reading: **Periodic Tales**

Carvone up the smell

How can two carvone molecules with identical chemical formulae and the same physical properties smell totally different?

Answer overleaf ➜

While you're thinking ...

Carvone is a relatively simple aromatic (meaning it contains a hexagonal ring of carbon atoms) organic compound with the formula $C_{10}H_{14}O$.

This interesting smelly compound is a member of the terpene family, a substance found in plant resins, which makes up a major component of turpentine (the name terpene comes from the same root, as turpentine was originally spelled 'terpentin').

Carvone is primarily found in the oil of two plants, giving them their entirely different, distinctive odours.

Because the molecules are chiral – the same structure, but not the same as their mirror image

Specifically, the two versions of the carvone molecules are optical isomers, also known as enantiomers, meaning that they are non-identical mirror images of each other – so score a point for any of the three. Some objects are identical to their mirror image. But others are different, making it impossible to superimpose one on the other. The usual example is a hand or a glove – there are clear, different left- and right-hand variants. The same holds for carvone molecules. One form, known as S-(+)-carvone gives caraway seeds their distinctive smell, while the mirror image form, R-(–)-carvone, provides the sweet and minty odour of spearmint.

Exactly how this subtle difference in arrangement of the molecules is detected by the nose has been subject to lengthy debate. It was originally widely thought that it was an effect of the shape of the proteins in the nose that lock on to different 'smelly' molecules. But as we have gained more information about the workings of the nose at the molecular level, this seems unlikely – molecules with very similar shapes can smell very different and vice versa.

It has been suggested that this may be a quantum process where the receptor can somehow detect the shape using tunnelling effects, but a more likely suggestion is that pairs of enantiomers that smell the same tend to have rigid carbon rings, while those that smell different are flexible. This may mean that the ability to detect a specific smell includes some kind of physical manipulation of the molecule. But the jury is still out.

*Further reading: **Royal Society of Chemistry** – The Crucible*

The colour purple

Why did mauve become a far more popular colour in fashion after the 1850s?

Answer overleaf ➜

While you're thinking ...

Purple dyes have been used since ancient times, but were very uncommon until the 1850s, when they suddenly took off in a big way.

The word 'mauve' is relatively new, first seen in the late 1790s and generally used for a light purple – somewhere between violet and lilac; the name derives from the French word for the mallow flower.

Like a number of colour names, 'purple' has shifted over the years. Originally the word was used to describe pretty well any shade of red (sometimes used to describe the colour of blood). But over time it became specifically identified with a colour that had been called Tyrian purple, containing more blue, that comes between crimson and violet.

Because of the introduction of synthetic dyes

The Tyrian purple dye is produced from a secretion of a range of sea snails found in the Mediterranean. It can be obtained by either irritating the snails, causing them to secrete the substance, or crushing them up. Although the first approach is more sustainable, it produces significantly less dye, so traditionally the snails were more likely to be destroyed.

It took thousands of snails to produce enough dye to deal with a single piece of cloth, which meant that the dye was very expensive. Originally used by the Phoenicians as long as 3,500 years ago, it was the Romans who established the dye as the definitive colour for emperors, senators and others of high office. With no obvious alternative to produce lasting dyes in this colour range, purple remained a rare shade for garments all the way up to Victorian times.

However, in 1856, the English chemist William Henry Perkin produced the first of a range of synthetic dyes that would be known as aniline dyes. His first product would be known as mauveine, and was discovered by accident. Perkin was trying to make a synthetic version of quinine, at the time expensive as it was the only effective treatment for the symptoms of malaria. He treated the readily available organic compound aniline with potassium dichromate, but impurities in the chemical resulted in a black substance forming. When Perkin tried to clean it out, it produced a strong purple colour, which proved highly effective as a dye. Mauveine is, in fact, not a single molecule, but a mix of four related chemicals. It was originally called aniline purple, but by the end of the decade the colour had become mauve and the substance mauveine. Soon a whole range of aniline dyes would be developed, but mauve led the way.

Further reading: **Bright Earth**

Elementary lighting

Which two elements were combined to form the name of a lighting company?

Answer overleaf ➜

While you're thinking ...

One of the elements in question was singled out for its smelliness, particularly in certain compounds.

The name of the second element comes from the Swedish words for 'heavy stone', though oddly, the element has a different name in Swedish.

The lighting company was founded in 1919 in Germany, although its name was first used for products thirteen years earlier. It is still a major player in the lighting market today.

Osmium and tungsten – Osram

The element osmium was discovered in 1803 by the English chemist Smithson Tennant, along with another element, iridium. While iridium stood out for the way that it had compounds with a colourful, iridescent sheen, osmium's compounds were often unpleasantly pungent, and even the element tends to oxidise to produce the extremely odorous osmium tetroxide, so it got its name from the Greek word for smell, *osme*. The other claim to fame for osmium was its heaviness. Along with iridium it is one of the two densest of all the elements (there is still a degree of dispute over which wins). Osmium also has an extremely high melting point.

Tungsten had been discovered a few years earlier, in the 1780s. It too is dense and has a very high melting point – in fact, the highest of all the metals. It was their reluctance to melt that meant that osmium and tungsten were the metals of choice for the filament for the old type of incandescent light bulbs. Although tungsten was the most common material to be used, osmium was a common alternative in the early days, resulting in one of the best-known lighting manufacturers taking its name from a combination of the two elements: Osram.

Getting 'Os' from osmium is straightforward, but it's less obvious how you get 'ram' from tungsten, until you discover that tungsten's chemical symbol is W, which stands for wolfram, named after the mineral wolframite. This was the name for the element in many European countries, though tungsten is now the official chemical name throughout the world.

Further reading: **Periodic Tales**

QUIZ 2
FIRST SPECIAL ROUND: DAISY CHAIN

Each of the biological names below links to another, either by biological relationship or by a similarity of name. Insert the names into the table below so that they form a chain.

Vulpes vulpes   Aesculus hippocastanum   Lathyrus odoratus

Foxglove   Sweet chestnut   Groundnut

Elderflower   Digitalis trojana   Goutweed

One point for each correct connection and a bonus two points if you get them all correct.

Vulpes vulpes
Elderflower

Daisy chain – solution

Vulpes vulpes (fox)
Foxglove
Digitalis trojana (a type of foxglove/Trojan horse)
Aesculus hippocastanum (horse chestnut)
Sweet chestnut
Lathyrus odoratus (sweet pea)
Groundnut (peanut)
Goutweed (ground elder)
Elderflower

One point for each correct link with a bonus two for getting all eight.

QUIZ 2

SECOND SPECIAL ROUND: UNREAL SCIENTISTS

Identify the movie featuring these scientists from their photograph or description:

1.

2.

3.

4.

5.

Quiz 2 | Second Special Round: Unreal Scientists

6. Dr Grace Augustine wants to study the Na'vi, an intelligent alien race, but constantly battles the military and business interests.

7. Mathematician Ian Malcolm is an expert in chaos theory.

8. Botanist Mark Watney gets stranded and has to battle with the environment to survive.

9. Professor Barnhardt meets the humanoid alien Klaatu when his flying saucer lands in Washington.

10. Billionaire computer scientist Nathan Bateman invites one of his employees, Caleb Smith, to his remote home to take part in a Turing test.

Unreal scientists – solution

1. *Dr. Strangelove*

2. *Contact*

3. *First Men in the Moon*

4. *Forbidden Planet*

5. *Daleks – Invasion Earth: 2150 A.D.* (half a point for *Dr Who and the Daleks*)

6. *Avatar*

7. *Jurassic Park*

8. *The Martian*

9. *The Day the Earth Stood Still*

10. *Ex Machina*

FURTHER READING

If one of our topics catches your interest, here's a chance to find out more. Note that these books or articles are not necessarily the source of the information in the quiz, but will allow you to read further around the topic.

A Brief History of Infinity: The Quest to Think the Unthinkable, Brian Clegg (Constable & Robinson, 2003)

A History of Mathematics, Carl B. Boyer (John Wiley, 1991)

Are Numbers Real?, Brian Clegg (St Martin's Press, 2016)

Armageddon Science, Brian Clegg (St Martin's Press, 2010)

BBC Future – Asparagus, Claudia Hammond (BBC Future 2014: http://www.bbc.com/future/story/20140818-mystery-of -asparagus-and-urine)

Bright Earth: The Invention of Colour, Philip Ball (Penguin, 2002)

Dice World: Science and Life in a Random Universe, Brian Clegg (Icon Books, 2013)

Electronic Dreams, Tom Lean (Bloomsbury, 2016)

Elephants on Acid, Alex Boese (Pan Books 2009)

Extra Sensory, Brian Clegg (St Martin's Press, 2013)

Final Frontier, Brian Clegg (St Martin's Press, 2014)

Flatland, Edwin Abbott (Dover Publications, 1992)

Gravity's Engines, Caleb Scharf (Allen Lane, 2012)

Inflight Science: A Guide to the World from Your Airplane Window, Brian Clegg (Icon Books, 2011)

Instant Brainpower, Brian Clegg (Kogan Page, 1999)

Invertebrates, Richard Brusca and Gary Brusca (Sinauer Associates, 2003)

Learning in Fishes: from three-second memory to culture, Keven Laland et al. *Fish and Fisheries*, 2003, 4, 199–202

Light Years: The Extraordinary Story of Mankind's Fascination With Light, Brian Clegg (Icon Books, 2015)

Masters of Doom, David Kushner (Piatkus, 2003)

Measure for Measure, Alex Hebra (The John Hopkins University Press, 2003)

Nature Wrinkly, Becky Summers (Nature News 2013: http://www.nature.com/news/science-gets-a-grip-on-wrinkly -fingers-1.12175)

Periodic Tales, Hugh Aldersey-Williams (Penguin, 2012)

Physics for Gearheads, Randy Beikmann (Bentley Publishers, 2015)

Royal Society of Chemistry – Compounds various authors (www. rsc.org/chemistryworld/more/?type=podcasts-compounds)

Royal Society of Chemistry – The Crucible Philip Ball (www.rsc.org/ chemistryworld/Issues/2009/February/ColumnThecrucible.asp)

Schrödinger's Kittens and the Search for Reality, John Gribbin (Phoenix, 2003)

Science: A History, John Gribbin (Penguin, 2003)

Scott video, www.youtube.com/watch?v=4mTsrRZEMwA

Ten Billion Tomorrows: How Science Fiction Technology Became Reality and Shapes the Future, Brian Clegg (St Martin's Press, 2015)

The Cambridge Illustrated History of Astronomy, Michael Hoskin (Cambridge University Press, 1996)

The Collins Field Guide to Insects, Michael Chinery (Collins, 1993)

The God Effect, Brian Clegg (St Martin's Griffin, 2005)

The God Particle, Leon Lederman with Dick Teresi (Dell Publishing, 1993)

Reality's Frame: Relativity and Our Place in the Universe, Brian Clegg (Icon Books, 2017)

The History of Clocks and Watches, Eric Bruton (Grange Books, 2002)

The Library of Isaac Newton, John Harrison (Cambridge University Press, 2009)

The Magic of Maths, Arthur Benjamin (Basic Books, 2015)

The Mechanical Turk, Tom Standage (Allen Lane, 2002)

The Quantum Age: How the Physics of the Very Small Has Transformed Our Lives, Brian Clegg (Icon Books, 2014)

The Restaurant at the End of the Universe, Douglas Adams (Pan Books, 1980)

The Tyrannosaur Chronicles, David Hone (Bloomsbury Sigma, 2016)

The Universe Inside You: The Extreme Science of the Human Body, Brian Clegg (Icon Books, 2012)

Things to Make and Do in the Fourth Dimension, Matt Parker (Particular Books, 2014)

Universe Inside You website, Brian Clegg, 2012: www.universeinsideyou.com/experiment4.html

Venoms of the Hymenoptera, Tom Piek (Ed.) (Elsevier, 2013)

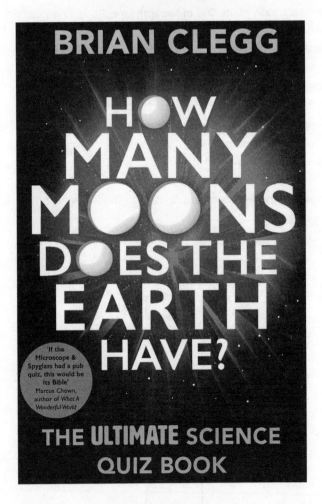